BY NOEL T. HOLLEY

November 4, 1988

N.J. International Inc.

1987

ISBN 0-934088-14-4
Library of Congress 87-061212

Published by: ***N.J. International Inc.***
77 West Nicholai St., Hicksville, N.Y. 11801

No. 7103

Printed in Hong Kong

THE MILWAUKEE ELECTRICS

An inside look at locomotives and railroading.

CHAPTERS

I. THE ROUTE TO PUGET SOUND 1

II. THE GE MOTORS 33

III. THE STEEPLECAB SWITCHERS 78

IV. THE BIPOLARS 86

V. THE WESTINGHOUSE MOTORS 102

VI. THE LITTLE JOES 118

VII. THE ROTARIES 146

VIII. THE ELECTRICAL SYSTEM 154

IX. THE IDAHO DIVISION 170

X. ECONOMICS AND COMPANY POLICY 210

XI. TALES OF THE RAILS 225

XII. TECHNICAL APPENDIX, LOCOMOTIVES 279

DEDICATION

This book is dedicated to all of those who love locomotives, railroading and the Milwaukee Road's west end. It is dedicated to the employees who proudly shared their memories, the photographers who recorded on film what they saw and the railfan/historians who shared the carefully collected records of the past. It is also dedicated to all of those who wish they could have been at trackside on the route of the electrified Olympian, but were never there. It is my hope that all of those to whom this is dedicated will find it a satisfying book.

ACKNOWLEDGEMENTS

Writing a book is an enormous project which can easily take years of spare time and vast amounts of energy. When I began, I estimated I would need three months. At the project's completion, I found it had taken four years and the help of seventy other people. A large part of what made it all worthwhile were those other people and the generous contributions of time and effort which they made. All are deserving of recognition and I have listed them here alphabetically.

Among the group, four deserve special note. These are Laurence Wylie, George Frazier, Adam Gratz and Pat Chester.

The late Laurence Wylie granted numerous interviews during which he shared the workings of the Milwaukee and its electrification. In addition, he loaned his negatives so that I could print those of his pictures which appear in the book. Wylie was an enthusiastic proponent of electrification and at the same time, a practical realist. His accomplishments were unquestionably worthy of note and I will be eternally grateful to have met him.

George Frazier was my earliest contact within the Milwaukee Road and over a period of years, he proved to be a generous and valuable source of technical information.

Adam Gratz and Pat Chester granted extensive interviews describing events in their careers on the railroad. Their insights helped tremendously to provide an understanding of railroading from the workingman's point of view. In addition, they referred me to many other retired employees whose memories added greatly to the book.

PHOTO CONTRIBUTIONS

C.R. "Casey" Adams
Ted Benson
Bruce Black
Alan Burns
Bruce Butler
Doug E. Cummings
The Denver Public Library
Dick Dorn
Electro-Motive Division
Albert Farrow
FEPASA Railway
Adam Gratz
Jim Fredrickson
General Electric
Philip R. Hastings
Gil Hulin
John Illman
Art Jacobsen
Donald R. Kaplan
Barry Kirk
Fred Kirk
Blair Kooistra
Russ Luedke
Warren R. McGee
The Milwaukee Public Library
The Milwaukee Road
Mrs. Harry Morgan
Ronald V. Nixon
Norfolk & Western Ry.
Jack Powers
Thomas Radoman
Larry Richards
Gordon Rogers
Richard Steinheimer
Stan Styles
Wally Swanson
C.L. "Bud" Tilbury
Harold K. Vollrath
Washington State Historical Soc.
Robert Webber
Warren Wing
Laurence Wylie
D. Larry Zeutschel
Rick A. Zeutschel

MILWAUKEE ROAD EMPLOYEES WHO CONTRIBUTED HISTORICAL INFORMATION AND STORIES

Bill Bibby
Alan Burns
Eugene R. "Wimpy" Burns
Pat Chester
King Clover
Ralph Danley
Paul Edmunds
Ralph Edwards
George Frazier
Walter Gordon
Adam Gratz
Hans M. "Swede" Hansen
T. Barry Kirk
Fred Kirk
Bill Lintz
Bill Merrill
Bill Plybon
Dean Radabaugh
Andrew Rogerson
Jim Scribbens
Wade Stevenson
Bill Wilkerson
Bob Williams
Don Wylie
Laurence Wylie
Frank Zellerhoff

THOSE WHO PROVIDED EXTENSIVE EDITORIAL ASSISTANCE AND RESEARCH ASSISTANCE

Don Dietrich
Art Jacobsen
Ed Lynch
Keith Newsom
Jack Powers
Larry Richards
Shelley Stockham
Warren Wing

THE ROUTE TO PUGET SOUND

The Milwaukee Road began rail operations on February 25, 1851, as the Milwaukee and Mississippi Railroad. The first official trip was a twenty mile run from Milwaukee to Waukesha, Wisconsin. During the next fifty years, it grew rapidly as a result of new contruction and the acquisition of other railroads. By the turn of the century, the railroad which was known as the Chicago, Milwaukee and St. Paul had 6,000 miles of track. It was a prosperous and independent midwestern farm belt railroad which had main offices in Chicago and ties to the famous Rockefeller family.

During the preceding twenty years both the Northern Pacific and the Great Northern had completed transcontinental lines to the Pacific Northwest. Their arrival broke the domination of the Northwest by the Union Pacific and the Harriman interests. Economic battles for routes, power and corporate influence ensued.

The "St. Paul," as the Milwaukee was then called, feared that without its own route to the West the future did not bode well. To the directors, it appeared that dependence on the U.P., N.P., or G.N. for a western connection would eventually result in their railroad being taken over or strangled. For protection, it was proposed that a Pacific extension be constructed jointly with another Midwestern railroad, the Chicago and Northwestern. When negotiations failed, the St. Paul pressed on alone. In 1906, construction began on a line which was to cross five mountain ranges on a route shorter and faster than those of the St. Paul's competitors. It was completed in 1909 without the benefit of land grants and at a cost which was more than five times the original engineering estimate. The St. Paul first began studying electrification of its mountain crossings in 1907 and let the first contract in 1914.

The high costs of construction, electrification, and the purchase of other Midwestern railroads drove the St. Paul into bankruptcy in 1925. It was reorganized in 1927 as the Chicago, Milwaukee, St. Paul and Pacific and thereafter known as The Milwaukee Road.

From its construction to its abandonment in 1980, the Milwaukee's Pacific Extension was a low traffic single track line operated with train orders and Automatic Block Signals. Centralized Traffic Control was never installed. There were few major improvements such as line relocations over the years and no major upgrading of facilities. Additionally, the track was lightly maintained compared to the Union Pacific, the Northern Pacific and the Great Northern.

In order to convey a clear understanding of this line which was the last of the transcontinentals, a map and descriptions of important points are provided.

CITIES AND TOWNS ALONG THE MAINLINE

TACOMA, WASHINGTON

Tacoma was the western terminus of the Coast Division electrification. It is an industrial port on Puget Sound and for many years it was the second largest city in the State of Washington. The Milwaukee's Tacoma Yard was built on a rip-rap fill in the tideflats. The yard complex included several company owned piers for ocean-going steamships, a car float dock so that freight cars could be barged to saw mills across the Sound, and a log dump for the Milwaukee's log trains. The yard itself was a large one and within the yard was the Milwaukee's best equipped shop west of Milwaukee, Wisconsin. In addition to a roundhouse and engine servicing facilities, there was a long brick shed for running repairs, a tall building with a traveling crane capable of lifting entire locomotives, a back shop served by a transfer table, a machine shop, an electrical shop and a car shop. Tacoma was the maintenance base for all the Coast Division steam, diesel and electric locomotives. It also handled most overhauls for Idaho division locomotives. Numerous branchline locomotives were commonly found in Tacoma along with the mainline power. Switching was generally performed by idle branchline locomotives. The only yard tracks electrified were those needed to allow road locomotives to get to and from their trains.

The Milwaukee did not enter Union Station which was served by the Union Pacific, the Northern Pacific, and the Great Northern. Instead it had its own small depot approximately a mile south. That old wooden depot was replaced in 1954 by a new brick one located in Tacoma Yard adjacent to the shops.

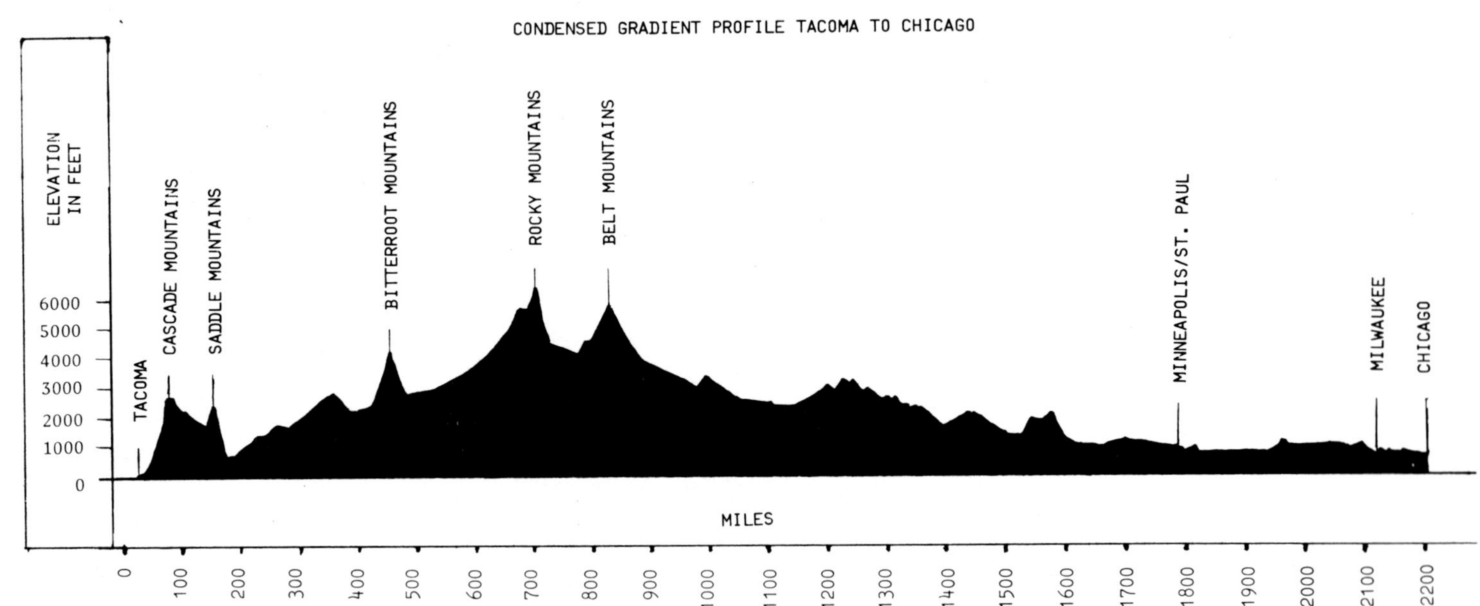

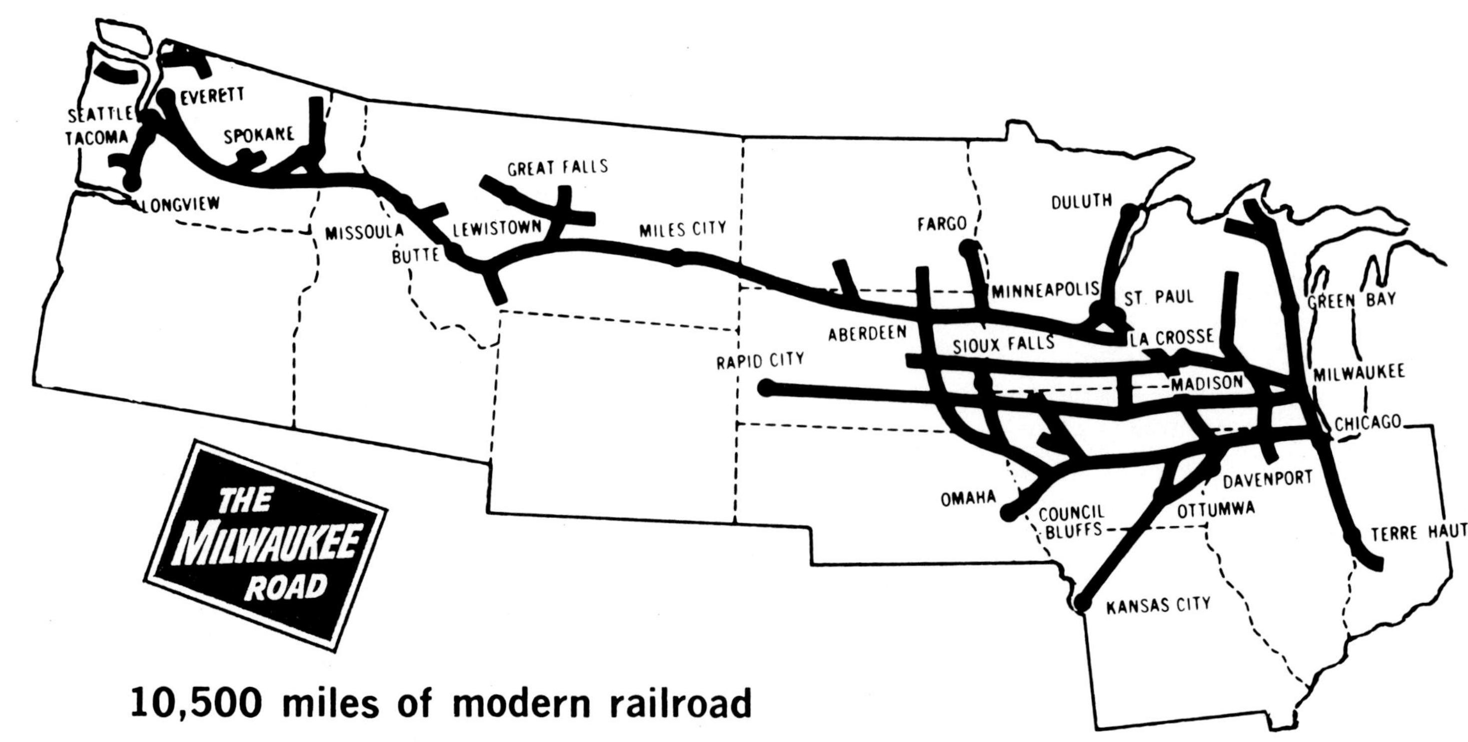

10,500 miles of modern railroad

SEATTLE, WASHINGTON

Seattle is located on the shores of Puget Sound. It is the largest city in the State of Washington and is a major seaport. It was home for the Milwaukee's western regional offices, and the Electrification Department was located in Seattle until the mid-1960's when it moved to Tacoma.

There were no major Milwaukee Road facilities in Seattle. It was already blanketed with rail trackage by the time the Milwaukee reached Puget Sound. Thus, the Milwaukee entered Seattle on the tracks of the Pacific Coast Railroad and the Union Pacific. The Milwaukee was a tenant in the U.P. owned passenger depot. It had one small train yard at Van Asselt, adjacent to the Union Pacific yard. A small switching yard and engine house was located four miles away at Stacy Street, adjacent to the Pacific Coast, and Northern Pacific yards.

On the waterfront the Milwaukee had a car float dock for barge service to its lines in Port Townsend and Bellingham. It also had access to the joint use Port of Seattle trackage. Electric locomotives operated to and from Seattle, but seldom spent time there. The Seattle locomotives were one or two switch engines.

THE MILWAUKEE SKI BOWL

The Ski Bowl was located at Hyak, the summit of the Milwaukee's line through Snoqualmie Pass. With a base elevation of 2,582 feet, it got enough Cascade Mountain snow to operate from January through March. Facilities included a large day lodge with pedestrian snow sheds to trackside, an electrically hauled sled type ski lift, rope tows, ski jumps and ski runs equipped with electric lights. Skiers were brought in by train from Seattle and Tacoma. The area opened in 1937. It was closed in 1949 after the lodge was destroyed by fire.

OTHELLO, WASHINGTON

Located on the flat arid lands of eastern Washington, this farming community would be surrounded by desert if it was not for irrigation. Othello yard was fairly large by Milwaukee standards, but because of less than ideal layout, it was sometimes a bottleneck to traffic.

Othello was the eastern terminus for the Coast Division electrification. Here the steam and diesel power used on the Idaho Division were replaced by electrics. Othello was the main shop for the Idaho division, but no heavy work was performed here. Locomotives requiring extensive overhauls were sent to Tacoma or Miles City. Sometimes the Othello shop changed out flues, superheaters and driver tires on steam locomotives. However, most work was light maintenance, painting, or running repairs. The only work regularly performed on electrics was oiling and sanding. Generally road locomotives could be found here awaiting trains. Switchers were not generally found in the yard since trains were blocked by road crews using road locomotives. Any other necessary switching was performed by branchline steam engines and diesels. Not all tracks were electrified.

SPOKANE, WASHINGTON

Spokane is located in a broad valley. It was for many years Washington's third largest city and in recent years, it surpassed Tacoma to become second largest. Spokane was not a major point on the Milwaukee. At the time the line was built, Spokane was already dominated by the Northern Pacific, the Great Northern and the Union Pacific. The Milwaukee entered Spokane from the east via a branchline and Union Pacific trackage rights. The line through Spokane was often referred to as "the Passenger Line." Passenger trains used the Union Pacific depot.

The Milwaukee's original yard in Spokane was a narrow one built in a cut beneath the U.P. passenger depot. Engine facilities were a two stall shed located in the cut and a turntable at the end of the cut. This narrow yard could only hold 50 cars per track and by the 1950's, the Milwaukee had outgrown it. In 1954 a new, larger yard and engine facilities were constructed on the east side of Spokane. They were adjacent to the U.P. yard.

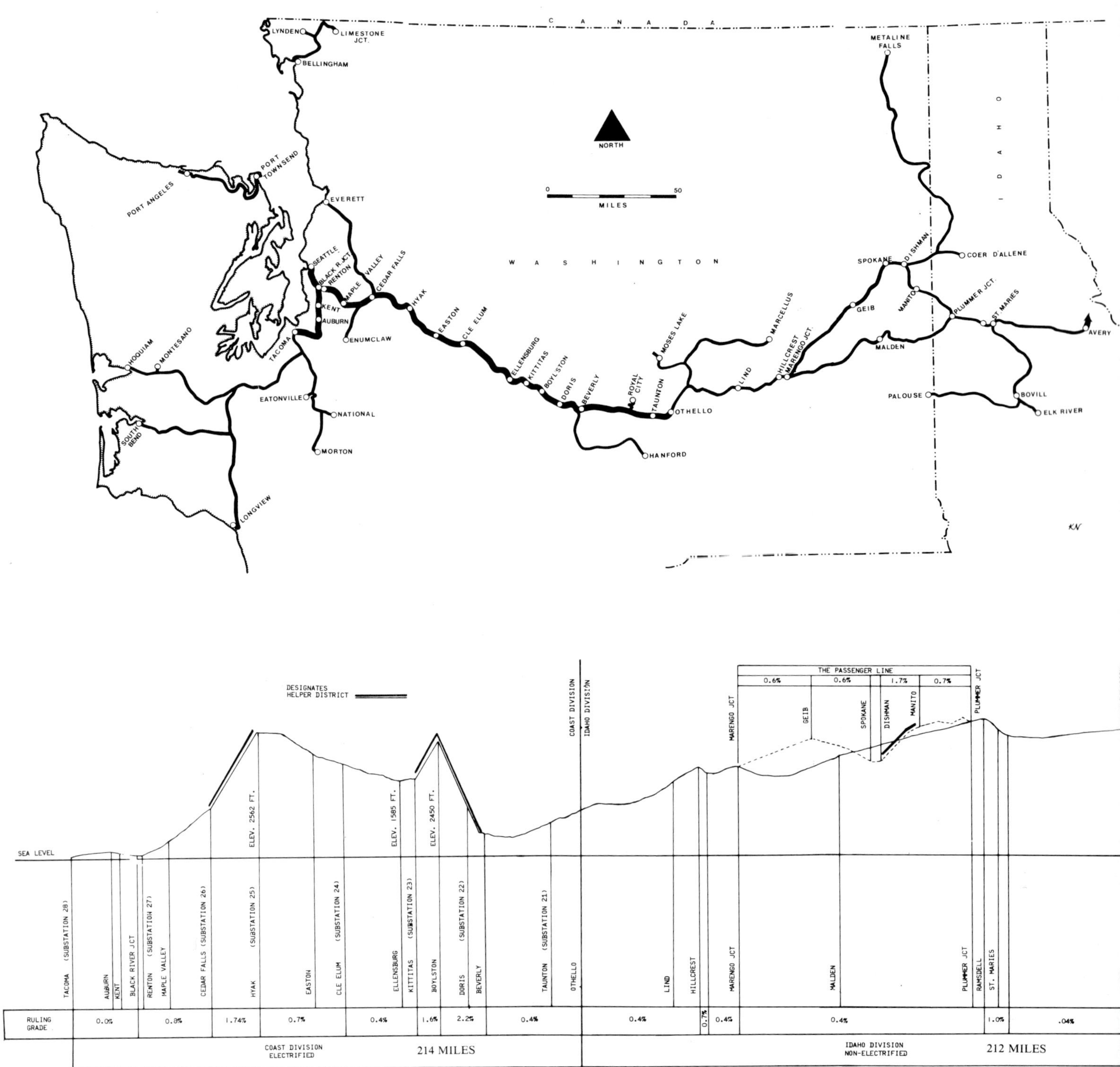

MALDEN, WASHINGTON

Malden is a tiny town located beside Pine Creek in the rolling pine dotted farm lands 30 miles south of Spokane. It was an Idaho Division crew change point and in steam days an engine change point. It was dominated by the railroad. It contained a roundhouse, dispatcher's office, a rip track, a beanery, and in its wilder days, four saloons and two whore houses. The line through Malden was entirely Milwaukee owned and was often referred to as "the Freight line."

AVERY, IDAHO

Avery was the division point between the Rocky Mountain Division and the Idaho Division. It was also the western terminus of the Rocky Mountain Division electrification. It is a small isolated community in a forested canyon and it sat at the base of the Milwaukee's 1.7% grade into the Bitterroot Mountains. Avery was connected with the outside world by the railroad and a one lane dirt road. Most of its residents were either railroaders or loggers. Railroaders who loved the woodsy out-of-doors loved Avery. The others considered it an American Siberia.

The importance of Avery to the railroad stemmed from the fact that it was an engine change terminal, a helper terminal and a crew change point.

The railroad yard here was long, narrow and built in an arc which almost formed an "S". It was bounded on one side by a mountain slope and on the other by the banks of the St. Joe River. It was common to find several locomotives, electric, steam or diesel near the roundhouse. Between trains they were serviced and given minor repairs. Since Avery is located at the edge of a zone of heavy snowfall and snowslides, one or two rotary plows were generally found in the roundhouse.

MISSOULA, MONTANA

Missoula is located in a broad valley and is one of Montana's largest cities. The only Milwaukee facilities located here were a large brick passenger depot and several sidings.

DEER LODGE, MONTANA

Deer Lodge was the site of a major switching yard and the Milwaukee's Rocky Mountain Division shops. It was located at the center of the Rocky Mountain Division in a broad western Montana valley, 226 miles from Harlowton and 211 miles from Avery.

Trains generally stopped here so that freight trains could be blocked and so that locomotives could be serviced. Locomotive servicing facilities here included a large roundhouse, a machine shop, an electrical shop and of course, fueling facilities providing oil and water.

The roundhouse was used for shop work on all types of motive power. It contained such equipment as drop pits, cranes and driver lathes. The machine shop was capable of manufacturing parts if need be. Even wreck damage could be repaired here, but major jobs were sometimes sent to the larger, better equipped shops at Tacoma, Washington, or Milwaukee, Wisconsin.

Generally several electric locomotives could be found on the Deer Lodge ready track, and several more inside the shop. An electric switcher worked the yard. A wreck train complete with big hook and a flat car load of 40 foot wooden poles was usually somewhere in the yard.

BUTTE, MONTANA

Butte is located in rolling hills at the base of the western slope of the Rocky Mountains. Although it is one of Montana's largest cities, the Milwaukee had nothing major located there. Butte Yard was a small one, primarily used for freight car setouts and pickups. A switcher was needed primarily for service to the industrial trackage in Butte. Helper locomotives were based in Butte Yard, but no servicing facilities were located there. If maintenance was necessary, the locomotives were sent to Deer Lodge.

A lineside landmark in Butte was the distinctive brick depot and clock tower which the Milwaukee built near the middle of town. This stub end depot was also used by the Butte Anaconda & Pacific. During the 1950's, the Milwaukee decided to eliminate the backing and wyeing of passenger trains. Thus, a small but modern depot was built along the mainline at the edge of town. The small building at Butte Yard was used only as a yard office and a train order station.

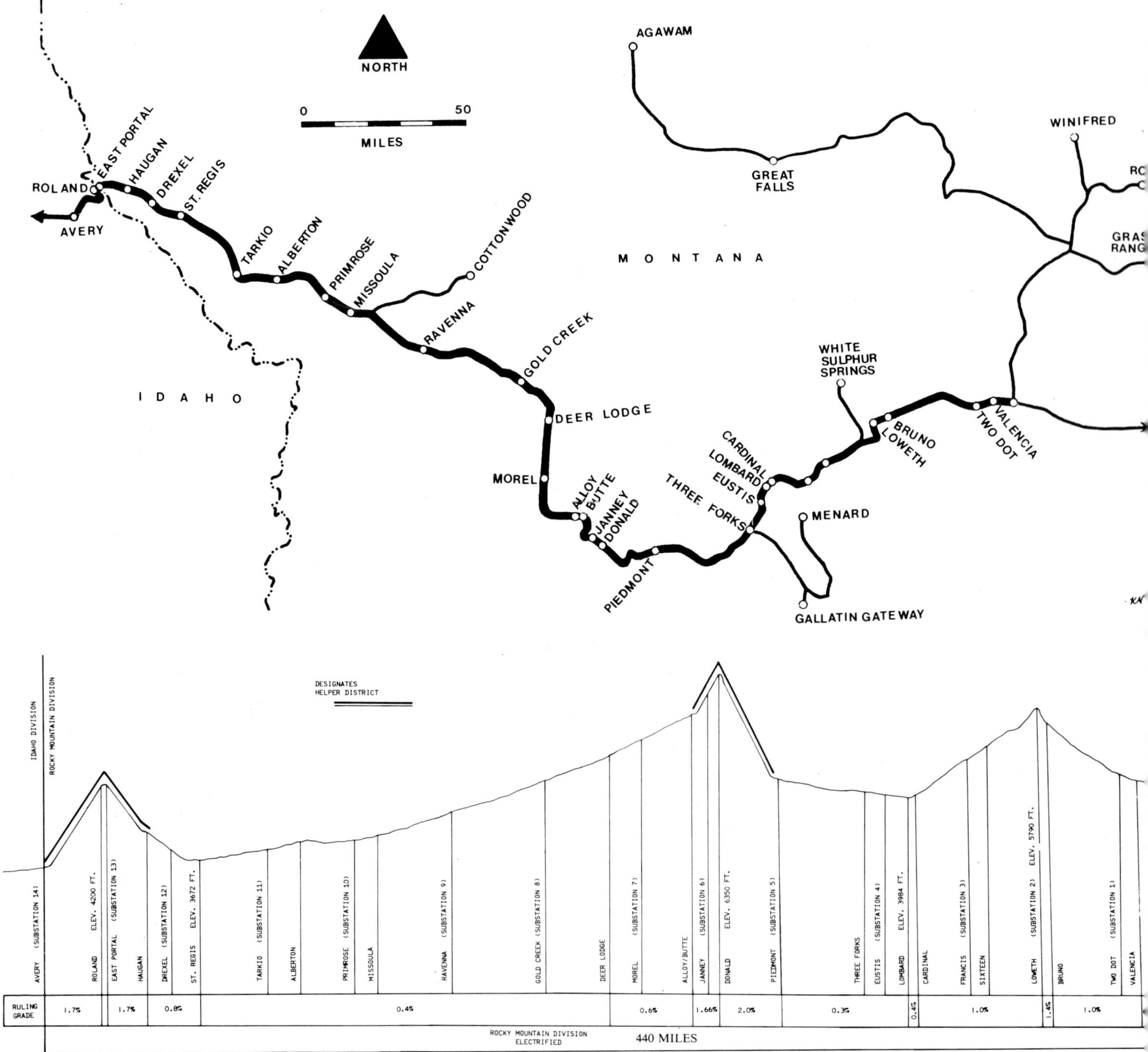

HARLOWTON, MONTANA

Located in the wide open spaces of the Big Sky State, the small town of Harlowton was the boundary of the Rocky Mountain Division and the Trans-Missouri Division. Harlowton was six miles east of the 1% grade which climbed into the Belt range of mountains. It was also the eastern terminus of the Rocky Mountain Division electrification.

Facilities in Harlowton were a large switching yard and an engine terminal. The terminal, which provided only fuel, water and running repairs served all types of motive power on a regular basis. There were mainline steam or diesels, from the Trans-Missouri Division, branch-line steam or diesels from the Great Falls branch and of course, the electrics. Generally, several locomotives could be found near the roundhouse and one or two electric switchers could be found in the yard.

CLE ELUM, WASHINGTON; ALBERTON, MONTANA; THREE FORKS, MONTANA

Prior to completion of the electrification, all three of these small towns were engine terminals with roundhouses. After steam engines were phased out, the roundhouses here were torn down. These towns retained small yards and depots. They continued to serve as freight service crew change points.

MILWAUKEE, WISCONSIN

Located on the shores of Lake Michigan, this Midwestern city was the sight of the Milwaukee Road's main shops. They were famed for building hundreds of steam locomotives, thousands of freight cars and fleets of streamlined passenger cars, in addition to performing major maintenance.

CHICAGO, ILLINOIS

Also located on the shores of Lake Michigan, Chicago is the business and financial center of the Midwest. The Milwaukee's corporate offices were located here, in Union Station, a building owned jointly by the Milwaukee and several other railroads.

HELPER DISTRICTS

Extra locomotives were often required to assist the movement of trains through the mountains. These "helper" locomotives were based and operated in districts where the mainline grades exceeded 1.5%. Prior to electrification, the helpers were extra steam engines coupled ahead of the road engine and also mallet pushers on the rear of the train. After electrification the helpers were GE Freight Motors operated in mid-train. Excessive concentrations of power at the ends of the trains tended to result in drawbars pulled out of freight cars and damage resulting from heavy slack action. Mid-train helpers running one-third to one-half of the way back in the train solved these problems. The helper was instructed to pull those cars behind it, but not to push those ahead of it. Helpers were valuable on descending grades as well as ascending grades because their braking power contributed substantially to train control.

THE CASCADE MOUNTAINS

The Cedar Falls Helper District was the 22 mile 1.74% grade from Cedar Falls eastbound to Hyak in Snoqualmie Pass. Hyak was the summit and helpers were generally removed from their trains there. The 0.7% grade westbound grade from Cle Elum did not require helpers. Base for the helpers was Cedar Falls. Generally, one or two GE Freight Motor sets were assigned there.

THE SADDLE MOUNTAINS

The Beverly Helper District consisted of a ten mile 1.6% eastbound grade from Kittitas to Boylston and an 18 mile 2.2% grade westbound from Beverly to Boylston. The 2.2% grade was the steepest mainline grade on the Milwaukee. Base for the helper locomotives was Beverly, Washington. Generally, one or two GE Freight Motor sets were assigned there.

THE SPOKANE VALLEY

The Mica Hill Helper District consisted of a seventeen mile 1.7% grade eastbound from Dishman to Manito. Base for the helper locomotives was Dishman, a point two miles east of Spokane. Helpers were, however, sometimes added to the trains in Spokane. Prior to dieselization, the helpers were usually 4-6-2's.

BITTERROOT MOUNTAINS

The Avery Helper District was the 24 mile 1.7% eastbound grade from Avery, Idaho, to East Portal, Montana, and the 14 mile 1.7% westbound grade from Haugen, Montana to East Portal. The line crossed the Bitterroots at St. Paul Pass. The crest of the Bitterroots is the boundary between the states of Idaho and Montana. Helper locomotives were based in Avery. Generally two GE Freight Motor sets were assigned there. At times, Idaho Division diesels which were in Avery awaiting trains were also used as helpers.

THE ROCKY MOUNTAINS

The Butte Helper District was the 15 mile 1.66% grade eastbound from Butte Yard to Donald and the 21 mile 2.0% grade from Piedmont to Donald. Donald was located on the continental divide at the summit of Pipestone Pass. The helper locomotives were based in Butte. Generally, two GE Freight Motor sets were assigned there.

THE BELT MOUNTAINS

The grades crossing the Belt Mountains were relatively mild, thus no helpers were regularly assigned there. The 44 mile eastbound grade was 1% from Lombard to the summit at Loweth. Most of the westbound grade was also 1%, but there was a short steep four mile section between Bruno and Loweth. This 2.1% grade sometimes required engineers to "double the hill." Cars were set out at Bruno so the road locomotives could take a light train to the summit. Then they would return and pick up the remaining cars. A line relocation in 1956 eased this grade to 1.4% and eliminated the need for doubling the hill. When eastbound trains were unusually heavy, helpers were sometimes used from Lombard to Loweth.

HELPER OPERATIONS PHASE-OUT

When electric operations were phased out, the Milwaukee also phased out its helper districts. This was accomplished by including extra diesels in the road locomotive consist on light trains and by using radio controlled mid-train slave units on heavier trains. The radio controlled diesels were operated in the train from terminal to terminal rather than only on grades.

Little Joes. E76 and E72 westbound with two GP-9s and a heavy train enter Avery Yard from the east./Photo by Richard Steinheimer.

E72, E74, 2058 and eastbound train #264 at Vendome, Montana on June 9, 1974. The Joe is in regeneration and the diesels are in dynamic braking as they descend the 2% grade out of Pipestone Pass in the Rockies./Photo by Dick Dorn.

The first drops of a summe thundershower splash into the oily soil of Deer Lodge Yard as drag freight #265 ties up after the day's run west from Harlowton. In summer, 1972, drags 265/266 were still 100% electrically powered, with Joes E71/E78 doing the honors on this Sunday July 16, 1972./Photo by Ted Benson.

"Pelican" E50 waiting at Haugan, Montana on a late afternoon in July of 1972./Photo by Ted Benson.

Bipolar 10252 snakes through Black River Yard showing the form that led some folks to call her a centipede./Photo by Asahel Curtis. Courtesy Washington State Historical Society.

Milwaukee Road Joes race a BN local westbound out of Deer Lodge, Montana...BN's short train length made them an easy winner! May, 1974./Photo by Ted Benson.

G.E. Freight Motor E39 ADCB, two GP-9s and an eastbound freight have just descended the 2.2% Saddle Mountain grade. Ahead lies the Columbia River Bridge and Beverly, Washington. Date 9-2-62./Photo by Doug Cummings.

June 9, 1974 Train #263 westbound at Ringling, Montana with E79, E71, 2060, 4008 and 354./Photo by Dick Dorn.

E73, E34 and E50 at the Avery engine terminal on a lazy August day in 1971. Avery, nestled at the foot of the Bitterroot Mountains, was the western end of the Rocky Mountain Division electrification./Photo by Noel T. Holley.

Train #263. The last westbound electric-powered train eases past the long closed "new" Butte depot on the morning of June 14, 1974. Power E79, E71, 3013, 3024 and 296./Photo by Dick Dorn.

Train #264 eastbound at Drummond, Montana with E72, 4005, 2058, 16 and 4008. June 13, 1974./Photo by Dick Dorn.

An eastbound freight has arrived in Avery. It has just crossed the non-electric Idaho Division and will soon assault the grades of the Rocky Mountain Division. 8-20-70./Photo by Jim Fredrickson.

General Electric diesels 5511, 5009 and 5004 are typical Idaho Division mainline motivepower in 1970, and they sometimes see use as Avery Helpers. They sit near the Avery roundhouse on 8-20-70./Photo by Jim Fredrickson.

Night trick at the Alberton, Montana depot. March 1973./Photo by Noel T. Holley.

Alberton, Montana depot. March 1973./Photo by Noel T. Holley.

Above:
Bipolar E1 entering Tacoma with eleven cars, 5-22-41./Photo by Albert Farrow.

The old Tacoma depot on "D" Street. Two passenger trains have arrived, 4-25-44. The large inspection car was built by Budda; the small one by Chevrolet./Photo by Albert Farrow.

Above:
A wheel inspection for Bipolar #10254 and Olympian, Train #16 at the Tacoma, Washington, depot. October 1, 1931./Photo by Otto Perry.

The "new" Tacoma passenger depot at Tacoma Yard in 1954./Photo courtesy The Milwaukee Road.

Bipolar 10253 and the Olympian head south out of Seattle past the Globe Feed Mill on 12-7-27. At this point, the 3,000 volt Milwaukee crossed at grade a line of the 550 volt Seattle Municipal Railway./Photo by Asahel Curtis. Courtesy Washington State Historical Society.

Westinghouse Motor 10307 and a passenger train prepare to depart the Butte, Montana depot. Date unknown./Photo courtesy of the Milwaukee Public Library.

Pacific Coast 2-8-0 #15 passes the Renton depot 9-14-51. The Milwaukee ran on the tracks on this coal hauling shortline between Maple Valley and Seattle./Photo by Albert Farrow.

On a summer afternoon in 1971, E50A/E47C/E50B are being prepared to pull a night train to Deer Lodge. The aging Freight Motor had recently arrived in Avery from the Coast Division and was in need of repairs./Photo by Noel T. Holley.

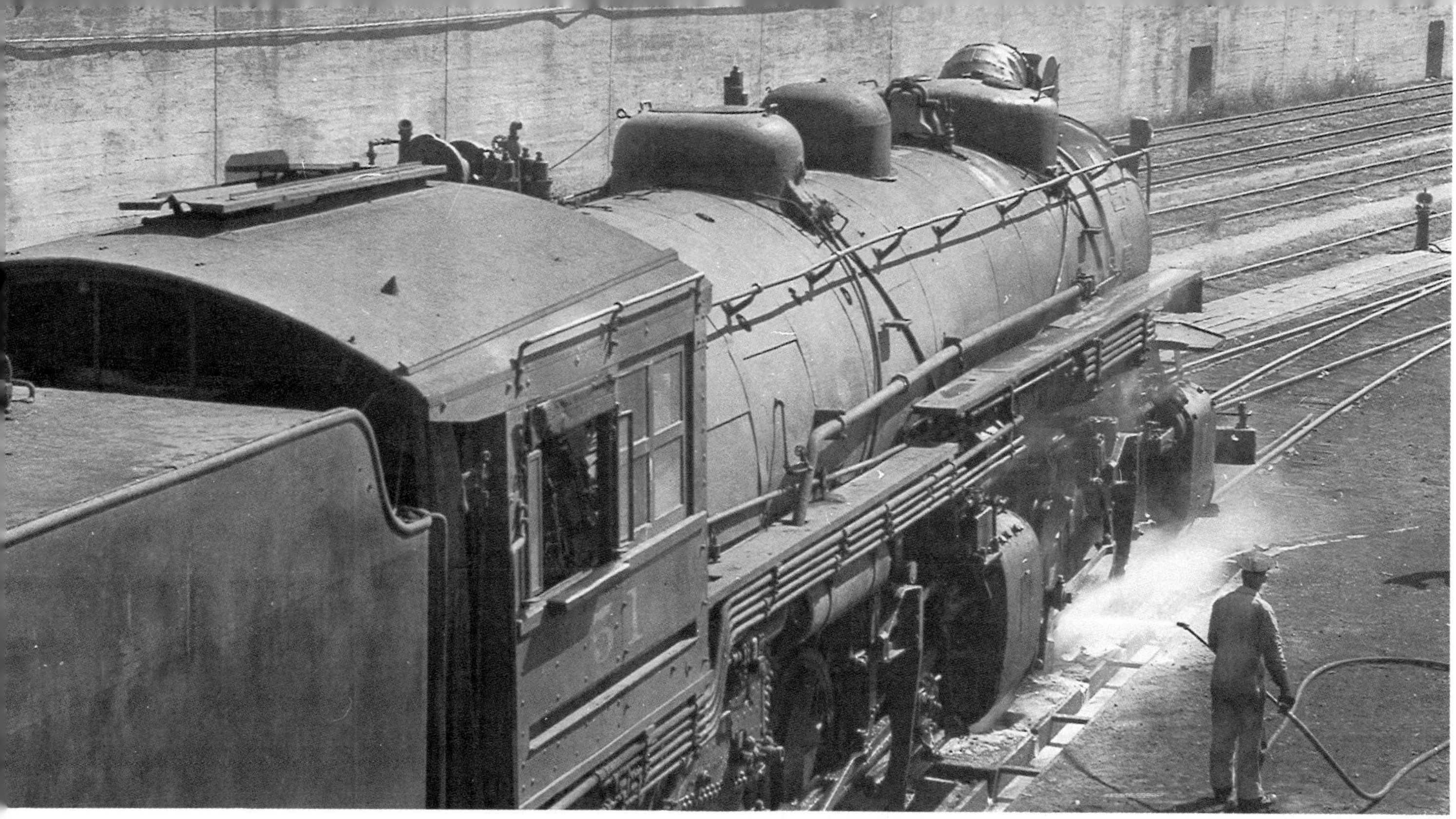

Mallet #51 gets a hot water wash near Spokane Union Station at the engine Terminal on June 25, 1951./Photo by Philip Hastings.

Train #261 rolls through Plummer Junction on the way to Othello. A string of general purpose GP-9 diesels is on the point. March 1964./Photo by Richard Steinheimer.

K Engine #2088 the Malden, Was ington roundhou in an early da photo. In 1938, t locomotive becar the 940./Photo R.V. Nixon.

Avery, Idaho engine terminal in 1915./Photo probably taken by Harold Theriault.

Little Joe E79 at Deer Lodge sh 4-9-60./Photo R.V. Nixon.

E-25A in the Deer Lodge roundhouse. Old as the hills, stronger than the mountain and too tough to die...almost. September 1973./Photo by Noel T. Holley.

"Goat" E81 receives maintenance repairs at the Deer Lodge shop. March 1973./Photo by Noel T. Holley.

Little Joe E-75 in the Deer Lodge roundhouse. September 1973./ Photo by Noel T. Holley.

Above:
Predecessor to the X3800 was the Cabless 10025. She was built on a truck from a wrecked Gallatin Valley streetcar. Because the electric locomotives rolled so easily, it took very little power to roll them in and out of the shops./Photo by Ron Nixon.

Above:
The Deer Lodge Shop Switcher, electric which ran on an extens cord pulls E77 in from the cold on October day in 1973./Photo Richard Steinheimer.

G.E. Freight Motor 10240 being rebuilt at the Tacoma shops following a wreck in the Cascades. Before salvage was begun, she had been branded a total loss./Photo by Laurence Wylie.

Freight Motor 10240 A & B after being rebuilt by the Tacoma shops. Also shown is the Tacoma shop complex. Date unknown./Photos by Laurence Wylie.

The 90, an N-2 class 2-6-6-2 is shown at the Tacoma shops. The front engine is undergoing repairs. 3-4-48./ Photo by Albert Farrow.

*September 13, 1963 finds E39 and E22 in the Tacoma engine house. They are being readied for eastbound freight runs./*Photo by Doug Cummings.

Dropping off the mountain into a white veil comes #263 leaving Roland on 5-15-74 behind Joes E20/E77...obviously the first train over the hill since the snow began, as the frosted rails mutely attest./Photo by Ted Benson.

Below:
Westinghouse Motor E17 and the second section of the westbound Olympian roll by Donald, Montana on March 17, 1946. Donald is the summit of the 2% grade in Pipestone Pass./Photo by W.R. McGee.

Left:
E5 and the ski train ha just arrived at the M waukee Ski Bowl. Da unknown./Photo by Johanson.

Right:
Nestled in the forest beneath the peaks of Snoqualmie Pass is the Milwaukee Ski Bowl Lodge. The Milwaukee mainline runs just behind the Lodge. It is reached through pedestrian snow sheds./Photo by Edwin Johanson.

Below:
The Milwaukee Ski Bowl at Hyak. The heavily rutted ski run is the landing area for ski jumpers. Date unknown./Photo by Edwin Johanson.

Above:
Typical skier cruising the slopes at the Milwaukee Ski Bowl./Photo by Edwin Johanson.

A steam rotary clears tracks at Hyak. It is pushed by a class N-3 mallet (2-6-6-2)./Photo by Edwin Johanson.

Crew change on a snowy October 31, 1973. The afternoon switch crew goes off duty./Photo by Richard Steinheimer.

The Mica Hill Helper, 4-6-2 #889 stands ready to help 4-8-4 #262 and the Olympian Hiawatha eastbound out of Spokane on the night of June 26, 1951./Photo by Philip R. Hastings.

Diesel #2002 A & B help 4-8-4 #250 and an extra westbound freight near Plummer Junction on May 27, 1951. The short 1% Plummer Hill grade was not designated as a helper district, but helpers were occasionally used there./Photo by Philip R. Hastings.

Freight Motor E42 ACB, the Cedar Falls Helper, drifts downgrade across Mine Creek trestle in the rain after assisting a heavy train to the summit of Snoqualmie Pass at Hyak. March 1962./Photo by Richard Steinheimer.

Above:
Arnold Hagemo at the throttle of E34 ACDB. The helper is running light to Avery after assisting a freight to East Portal. July 1970./Photo by Noel T. Holley.

Above right:
E45 ACB, the Avery Helper, is running mid-train on an eastbound freight. As the train crosses a trestle and enters a tunnel, I am braced tightly in my seat. Slack action in the train was enough to throw an unwary rider on the floor./Photo by Noel T. Holley.

Below:
June 10, 1974. Train #264 eastbound through Martinsdale, Montana with E77, 4055, 4006 and 2058. They are descending the Belt Mountain grade./Photo by Dick Dorn.

THE GE MOTORS
CLASSES EP-1, EF-1, EF-2, EF-3, EF-5, EP-1A and ES-3

These locomotives became the backbone of the Milwaukee's electric locomotive fleet, and some lasted an amazing 58 years in service. 84 units were constructed and they entered service as 42, two-unit locomotives. They were called by many names which included GE Motors, GE's, Freight Motors, Passenger Motors and Boxcabs. In certain localities railroaders called these slow but reliable locomotives Pelicans, Black Cows, and Mules. Although there was no agreement as to which names were proper, some railroaders considered the animal names objectionable. The company referred to these motors by class number and left the naming to others.

These locomotives were ordered from General Electric as part of a package deal. In 1914, GE offered to design a 3,000 volt DC electrification for the Milwaukee, to supply electrical equipment, and to supply locomotives under one large contract. The Milwaukee accepted GE's offer, and thus the work began.

Nineteen-fourteen was a time of beginnings and change. World War I began in Europe and it threatened a nation of isolationists on the opposite side of the Atlantic. The Panama Canal was opened and it threatened the Milwaukee's future as a land bridge to the Orient. Manned flight, though fledgling in nature was 11 years old and threatened no one since the railroads dominated land transportation by hauling most passengers and 75% of all freight.

In 1914 steam engines were the conventional motive power in heavy mainline service. Pacifics and Mikados were modern, and the Milwaukee's compound 2-6-6-2's could still legitimately be called mammoths of the rails.

What the General Electric designers proposed in 1914 was a radical departure from all of that. It was direct current locomotives which were huge compared to anything previously built: articulated locomotives with four, four-wheel driving trucks all linked by drawbars, a four wheel pilot truck on each end, and the ability to outpull and outrun anything on the Milwaukee's roster. GE utilized technology developed and proven in streetcars, refined on the Butte, Anaconda and Pacific, and pushed to new heights for the locomotives, which newspaper ads described as the "World's Mightiest."

In what was most likely an effort to speed production of the electric locomotive fleet, General Electric hired American Locomotive Company as a subcontractor. General Electric's Erie, Pennsylvania, plant built superstructures and running gear for 22 Freight Motors and electrical gear for the entire fleet. American Locomotive Company built superstructures and running gear for the 8 remaining Freight Motors and 12 Passenger Motors. ALCO's work was performed to GE specifications so that the appearance of all locomotives was identical. The first GE Motor was delivered in September of 1915, ten months after the contract was signed. The last arrived in January of 1917.

The 112 foot length of these locomotives reflected what GE called "the principle of stretching out the locomotive." The idea was to minimize the impact of vertical and horizontal blows to the track, by spreading the weight and mass of the locomotive over a large area.

Each locomotive was considered a unit consisting of two identical half-units. The half-units were, however, capable of independent operation. This utilitarian feature was designed to allow the Milwaukee to run half-unit locomotives in local service if it chose to. Each half-unit mounted a coupler on its front and a drawbar on its rear. The couplers and drawbars were, however, mounted in universal pockets which could accept either couplers or drawbars.

Double headed steam at Donald on Pipestone Pass prior to electrification./Photo from Thomas Radoman Collection.

The GE Motors were 14 feet 11¼ inches tall, 10 feet wide and 112 feet long. The Freight Motors weighed 288 tons and the Passenger Motors 317. Starting tractive effort at 30% adhesion was 135,000 pounds and 148,000 pounds respectively. Continuous horsepower at 3,000 volts was 3,000 and the one hour overload rating was 3,440 hp. In normal operation, each locomotive drew 840 amps. When called upon to produce 3,440 horsepower a locomotive drew 960 amps. If it required 135,000 pounds of tractive effort to start a train, the freight motors drew 2,920 amps. That amperage level could only be sustained for a brief matter of minutes without burning up the electrical equipment.

The locomotives had a 2-B + B + B + B-2 wheel arrangement. GE sometimes referred to it as a 4-4-4-4-4-4. Solid wheels 36 inches in diameter were used in the pilot trucks at each end. Spoked wheels 52 inches in diameter were in each driving truck. The driving trucks carried all pulling and pushing forces through their frames via articulated joints. Unlike latter day diesels, the box shaped cabs simply rode above the trucks. They were not a part of the locomotive frame. The total wheel base was 102 feet 8 inches, but the rigid wheel base was a very flexible 10 feet 6 inches. This short rigid wheel base made the GE Motors well suited to service on a mainline where ten degree curves were common.

There were 8 traction motors and each was huge. They were designed when electrical engineering was in its infancy and thus imprecise. Designers estimated the size needed to meet the performance specifications, then they added a large rule of thumb amount as a safety factor. The result was motors that could handle an overload of 171% for thirty minutes. Each traction motor was designed for 1,500 volts. They were wired to operate in pairs to allow operation at 3,000 volts. The nose of each motor was suspended from the locomotive chassis, while the other end was hung on a driving axle.

Each axle had two drive gears and each motor had two pinions. With this arrangement, the traction motor could safely operate even if gears or gear teeth broke on one side. The drive gears used were of a unique type called spring gears. Each one contained an outer ring, six coil springs, an inner ring which was attached to the axle, and 48 other parts. They were designed to absorb the jolts of sudden starts through compressing the springs. This, it was hoped, would prevent mechanical damage to the locomotives.

The gear ratio used on the Passenger Motors was 2.4 to 1, while that used on the Freight Motors was 4.5 to 1. As built, the top speed of the Passenger Motors was 70 mph; for the Freight Motors it was 35 mph.

Each half-unit had an operators cab at one end. Behind the cab wall an aisleway went down each side of the half-unit. Between the aisles sat the contactors, resistor grids, two air compressors, and a motor-generator set. The motor-generator set produced low voltage current for the locomotive control circuits. Within it, a 100 horsepower 3,000 volt motor drove a 7½ kilowatt 120 volt generator. Additional equipment on board passenger locomotives was an oil fired boiler, fuel and water tanks, and the train lighting equipment.

On top of each half-unit was a headlight, bell, six resistor bank flues and a pantograph. The pantograph was raised and held against the trolley wire by springs, but the tension on the springs was created by drawing them tight with air cylinders. For this reason the pantograph was classified as air-raised. On top of one half-unit in each set was a trolley pole called a "fishpole" or "stinger." The trolley pole was not used while operating the locomotive and no main circuit switches were to be closed while the pole was up. Its only purpose was to provide current for the air compressors, M.G. set and cab heater when the pantograph could not be raised. Due to inevitable air leaks, there were times when no air pressure was available for raising the pantographs. Without a back-up means of current collection there would be no way to start the air compressors.

Brand new Passenger Motor 10107 AB is shown being towed westbound through Milwaukee, Wisconsin. It was constructed in September 1916 and entered service soon afterward. Years later, it operated as Coast Division Freight Motor E33./Photo from Robert Weber Collection.

CLASS EP-1 THE PASSENGER MOTORS

The Milwaukee's twelve high geared GE Motors were numbered 10100 A & B through 10111 A & B. Crews just called them Passenger Motors. That was the only name they needed during the brief four years they rolled the coaches and Pullmans of the Olympian. They could not be confused with anything else because the only other electric locomotives running were the Freight Motors.

The first two locomotives, the 10100 A & B and 10101 A & B arrived in December of 1915. The remainder followed in short order with the last 10111 A & B arriving in January of 1917. They immediately took over passenger assignments between Harlowton and Avery, displacing steam power and two boiler equipped Freight Motors.

These fast, reliable motors were painted glossy black. They were trimmed with white lettering, and white striping on the pilot wheels. Externally, they appeared almost identical to the Freight Motors, but with higher gearing they ran faster. The Passenger Motors were capable of hauling only two-thirds of the tonnage Freight Motors hauled, but they could haul it twice as fast. This meant 60 to 70 mph through valleys, and 30 mph on mountain grades. According to retired Fireman Bill Merrill, the Milwaukee had little trouble with the Passenger Motors, but when starting trains they suffered from wheel slip due to weight transfer. The traction motors sat between the axles of each truck and they pointed in opposite directions. As the gears interacted during starting, the torque of one motor would work to lift its driver from the rails while the torque of the other pressed that driver down. Although the driver which was forced up by torque never left the rails, it tended to lose traction and spin.

In 1917, the Milwaukee's management decided to proceed with the second phase of mainline electrification. Construction crews went to work stringing wire between the Coast Division terminals of Othello and Tacoma. In order to operate the resulting 650 miles of electrified trackage, the Milwaukee estimated a need for a fleet totaling fifteen passenger locomotives and forty-two freight locomotives. Rather than ordering additional duplicates of the existing GE Motors, they decided to order a new state-of-the-art design. It was agreed that all of the new locomotives would be assigned to passenger service. The existing Passenger Motors would be regeared for freight service and assigned to the Coast Division. The 10100 A & B was converted in June of 1919 and the rest of the fleet followed during 1920 when the Westinghouse Motors entered service. The Passenger Motors then ceased to exist. They faded into the ranks of the GE Freight Motors, becoming 10230 A & B through 10241 A & B.

Passenger Motor Ten-One-Hundred AB pose for an Asahel Curtis photograph near the entrance of Silverbow Canyon./Photo courtesy of Milwaukee Public Library.

The Columbian ascends the 1.7% Bitterroot grade. The location is the Adair Loop and the date is approximately 1916. The slopes here were denuded by a fire in 1910./Photo by General Electric courtesy of Richard Steinheimer.

Freight Motor 10239 AB began its career as Passenger Motor 10109 AB. It was converted to freight service and sent to the Coast Division in 1920. This set was later known as E72 AB and E39 AB. This scene is on the Coast Division south of Black River. Date unknown./Photo by James A. Turner.

CLASS EF-1 THE FREIGHT MOTORS

Originally there were 30 electric locomotives in freight service and their road numbers were 10200 A & B through 10229 A & B. The number grew to 42 in 1920 with the conversion of the Passenger Motors.

The first of the group to arrive was the 10200 A & B, a locomotive which later became E50 A & B. It was delivered in September of 1915 and was the first of a group described by GE as "epoch making." They were painted black with white lettering like the steam locomotives, but in very short order they demonstrated their superiority over steam. They could haul more tonnage, they could haul it faster, their regenerative braking allowed safer control of trains on grades, they did not require water stops, and they did not require engine changes at 100 mile intervals. After the Freight Motors entered service, the engine terminals at Alberton and Three Forks, Montana became unnecessary. The roundhouses there were closed and the dollars saved helped to pay for electrification. Motive power was changed only at Harlowton, Deer Lodge and Avery. Another factor which impressed the Milwaukee was the availability of the GE Motors. There was very little on an electric which required servicing and so they never spent much time in the shops. Thirty Freight Motors could replace a much larger group of steam engines because steam, with its heavy servicing requirements, was not always available to pull trains.

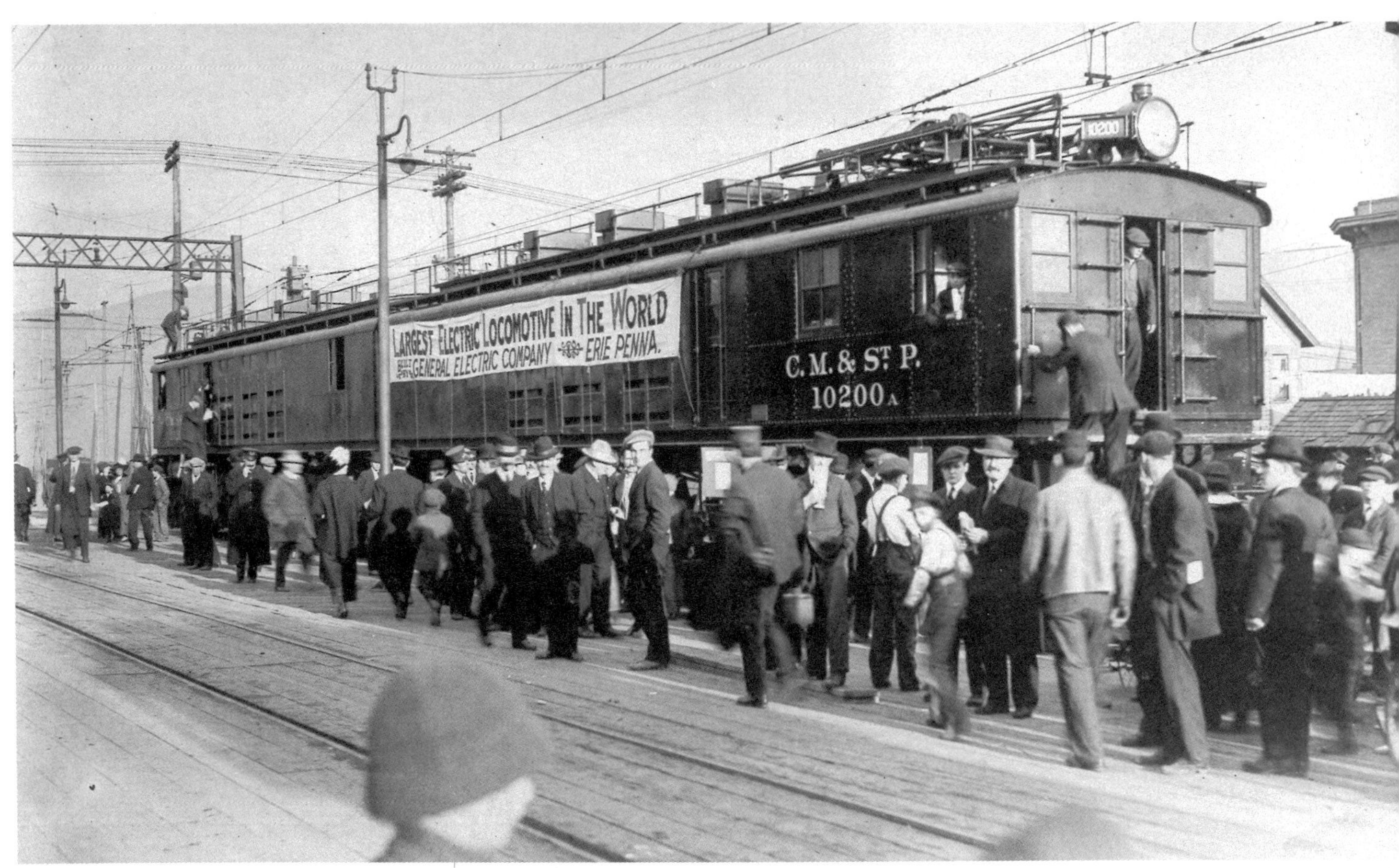

First of the Freight Motors, 10200 AB on display in Butte, Montana. Fireman Bill Merrill can be seen coming out the front door./Photo courtesy Milwaukee Road.

Above:
Freight Motors under construction at the General Electric plant in Erie, Pennsylvania./ Photo from Richard Steinheimer Collection.

An early Freight Motor rolls through a scene spliced together in the darkroom of Asahel Curtis. The location is Sixteen Mile Canyon and Eagle's Nest Tunnel July 21, 1916./Photo from General Electric.

In order to graphically illustrate the power and efficiency of electrics, for high company officials and the press, the Electrification Department staged a demonstration. A 3,000 ton freight train with the 10201 on the point and the 10200 as a pusher, attacked the western slope of the Rockies. Growling softly like streetcars, they averaged 16 mph on the 1.66% grade from Butte to the summit of Pipestone Pass. The speed and tonnage were amazing when compared to the typical 1915 freight that followed a 2,000 ton train with two thundering 2-8-2's on the point and a powerful 2-6-6-2 mallet pusher. That train was only able to achieve 9 mph. It was a sight which helped bring approval for completing the electrification of the Rocky Mountain Division and later the Coast.

From 1919 until 1932, the Freight Motors were operated only in A & B sets. Rainier Beeuwkes wanted it that way and Beeuwkes headed the electrification. He was appointed to his post in 1914 and held it through 1947. He was the Electrical Engineer and in his estimation the Milwaukee Road electrification was not only the longest in the world, it was the greatest. He felt that changes to the original designs would only serve to compromise excellence. Such tampering could bring unwanted problems.

Retired electrical engineering consultant Walter Gordon recalls the first time he proposed three-unit sets to Beeuwkes.

> "He wouldn't hear of it," says Walter. "Around Christmastime in 1927 I was working in the Electrification Department. I helped Beeuwkes and Laurence Wylie, who was one of his assistants, do a study of converting the Virginian electrification from AC to DC. AT & T was having problems with inductive interference caused by the Virginian, and our system didn't generally cause interference. During 1928, I took a trip to see the Virginian and one of the things that struck me was the thrcc-unit motors they ran. I said to myself why can't we do that, so I drew up a diagram and submitted it to Beeuwkes. We had just started running 4,500 ton trains the previous year, and these required a road locomotive and two helpers through the mountains. You can imagine the problems we had trying to get three engineers in different parts of the train to coordinate their activities. It was especially bad when they all tried to regenerate. What I proposed was running three-unit Boxcabs so that our freights would only need two locomotives instead of three. I recommended using MCB couplers too. They had built-in air lines and electrical connections so that you could quickly make up whatever size motor you needed. Beeuwkes wasn't interested, but he did leave the drawing in the files. In 1935, Wylie arranged for me to ride a private car across Montana and I saw a three-unit set in operation."

E64 AB nearing Auburn, Washington with a 38 car freight 6-7-46./Photo by Albert Farrow.

10225 AB sit by the Avery, Idaho roundhouse September 21, 1931. Note the pilot beam steps used on the Rocky Mountain Division. Pilot foot boards were used on the Coast./Photo by Otto Perry.

In September 1932, Coast Division Freight Motor 10234 AB became an EF-2. Its new number was 10500 ACB. In May of 1938, the long middle unit was replaced with a short one and the set became an EF-3. In 1939, the set was renumbered E25 ACB. Scene: Renton 1938 or 1939./Photo by James A. Turner.

CLASSES EF-2 AND EF-3

These were the three-unit Freight Motors, A's and B's with a third unit between them. Names for the middle units were simple and straight forward. They were "C units" or "middle units." The company term "half units" faded for obvious reasons. Such names as boosters, and slugs were invented by latter day railfans.

Three-unit Freight Motors were powerful, even if they were slow. Their horsepower output was 5,010 continuously and 6,150 for a one hour overload. They could haul 30% more tonnage up a 1.7% grade than could two 4-8-4's, and on a much easier .4% grade, they were rated at only 6% less tonnage than two 4-8-4's. All that they lacked was speed. The Freight Motors, whether EF-1's, EF-2's or EF-3's had a 35 mph speed limit. At higher speeds, centrifugal force might tear the rapidly revolving traction motor armatures apart.

The first EF-2's were assembled in September of 1932. They were given numbers 10500 ACB, 10501 ACB, and 10502 ACB. These sets were made by using 10234A, 10235A and 10235B as middle units. Pilots were removed, the front couplers were replaced with drawbars, and control circuit connections were added to the front ends.

No one still living recalls why Rainier Beeuwkes relented in his opposition to running three-unit Freight Motors. However, he was apparently pleased with the operation of the first group. Five more EF-2 motors followed in 1933, one in 1935, and three in 1936.

1936 also marked the construction of the first three EF-3 motors. These differed from the previous series in that the middle units had no control cabs or pilot trucks. The Deer Lodge shops constructed these units by removing the rear of the superstructure, plating over some unnecessary windows, and by cutting, welding and rebuilding the heavy bar frames of the front trucks. Deer Lodge shop crews called these new middle units "bobtails." The resulting EF-3 locomotives had 15,000 pounds more weight on the drivers and 2,500 pounds more tractive effort. This increase was, however, insignificant in the eyes of the train dispatchers. EF-2's and EF-3's were both assigned the same tonnage rating.

During 1936, three cabless middle units were built. They were followed by two in 1937, four in 1938, and three in 1939.

The process of building cabless units involved rolling a unit with one road number into the shops, rebuilding it, and then assigning it the road number of a C unit which had a cab. It then replaced that C unit in an existing three unit set. The original C unit became an A or B unit in a different set or was hauled into the shops for rebuilding and a new road number. Only twelve cabless units were ever built.

Although the Milwaukee continued to assemble new three-unit Freight Motor combinations between 1940 and 1954, all of them were EF-2's. The middle units had cabs. Retaining cabs allowed the Milwaukee greater freedom in deciding what the future use of each unit would be. Many of them alternated between being A's, B's or middle units in several different locomotives. The cabless units were middle units permanently.

E28C, a typical shortened middle unit, sits at Othello. Date unknown./Photo by Gayle Christian from the Jack Powers Collection.

Above:
10507 ACB begins the climb into the Cascades after passing Cedar Falls depot. The year is 1938 and the 10507 has been an EF-3 for one year./Photo by Albert Farrow.

Milwaukee Extra E-34 west in the hole for No. 16 between Three Forks and Harlowton. 5-29-50./ Photo by John C. Illman.

EF-3 Freight Motor E32 ACB near Seattle in 1952./Photo by Casey Adams.

EF-2 Freight Motor E38 ACB in the Alberton Yard, July 1957./Photo by Casey Adams.

CHANGING TIMES

The time period during which the EF-2's and EF-3's were first proposed and then assembled was one of evolution for motive power on many railroads. On the Milwaukee, however, it was a time of financial troubles. It was a time when the Chief Mechanical Officer, the ruler of steam, and the Electrical Engineer, the ruler of electrics, had to make do with the power they had.

The planners of the Milwaukee's Pacific Extension had envisioned high levels of traffic, but it never came. Thus, the Idaho Division was never electrified and the entire Pacific Extension sat underutilized. It had the capacity to handle ten times the traffic it carried.

In 1921 and 1922, the company purchased railroads in Indiana and Illinois to help generate coal and industrial traffic, but these lines required expensive upgrading. The cost of this plus the fact that constructing the Pacific Extension cost more than four times the budgeted amount, pushed the Milwaukee into bankruptcy. It filed in 1925, and due to the ravages of the Depression it filed again in 1935.

During the 1920's, motive power was evolving to new heights on other railroads, and the Milwaukee's major competitors upgraded. In 1926, the Northern Pacific bought its first twelve 4-8-4's. Those performed so well that between 1933 and 1943 thirty-seven more were purchased. Additionally, the N.P. bought forty-seven 4-6-6-4's from 1936-43. The Great Northern bought twenty 4-8-4's in 1929-30. The Great Northern's most rugged mountain crossing was in the Cascades, and in 1929, that line was modernized with an eight mile tunnel, a 73 mile electrification and eighteen new electric locomotives. The Great Northern's dual service Y-1 electrics were twice as fast as the Milwaukee's 1915 Freight Motors and twice as powerful on a unit-for-unit basis.

The bankrupt Milwaukee did not own a fleet of 4-8-4's until 1937 and no new electrics were purchased until the diesel era. In 1934, the Pennsylvania was building its legendary GG1's, but the Milwaukee was cash short. It could afford only to assemble EF-2's and EF-3's.

The next major step in the evolution of motive power was the introduction of mass produced freight diesels. Their arrival in the west served to emphasize even more the weaknesses of the Milwaukee's existing freight power.

In 1939, four unit FT diesels from Electro-Motive Division of General Motors toured the railroads as demonstrators. They took the Milwaukee by storm with impressive performances on the Idaho Division, across the electric zones, and everywhere else. Compared to steam they hauled more tonnage and hauled it faster. They ran past water stops and did not require frequent engine changes. Steam could not compete with them and the EF-2 and EF-3 Freight Motors could not match their speed and tonnage ratings.

In 1941, the Mechanical Department purchased one four-unit set of FT diesels. During World War II, the war Production Board alotted them only seven more sets, thus 4-8-4's were accepted. Dieselization began in earnest in 1949.

The Mechanical Department in Chicago looked forward to the prospect of modernization at long last. It also looked forward to ending its rivalry with the much smaller Electrification Department based in Seattle. The Mechanical Department had authority over all locomotives on the system except those powered by electricity. This fact rankled some mechanical engineers and created their desire to eliminate the electrification. The old and worn electric fleet was in need of replacement and the powerful new diesels looked just like the locomotives to do it.

In Seattle, Rainier Beeuwkes, the old man of the electrification, looked over his domain and resigned himself to its replacement by diesels. In addition to new locomotives, the electrification needed substations of greater capacity. A low traffic line such as the Milwaukee could not, he felt, afford that investment in the face of the economics available through dieselization.

In 1947, following the company's reorganization from bankruptcy, Charles Buford became company president. Buford was committed to modernizing operations and in 1947, he announced a ten-year plan for the replacement of all steam engines with diesels. The lines west of Harlowton were also tentatively slated for dieselization. In 1947, Beeuwkes was asked to retire from his post as head of the Electrification Department. J.P. Kiley, who was Buford's vice-president, appointed Wylie to the department head spot as of January 1, 1948.

Wylie came from the Operating Department where, in five years, he had worked his way up from Trainmaster to Division Superintendent. He was familiar with the electrification because he had worked in that department from 1915 until 1936. He earned a Masters Degree in electrical engineering from Montana State College in 1920 and worked as one of Beeuwkes' assistants from 1919 to 1936. Wylie had always aspired to head the Electrification Department but in 1936 he decided to transfer out rather than spending his career as an assistant to someone who would never retire. Upon appointment, Wylie's unofficial assignment was to develop an orderly phase-out plan for the electrification.

E30 ACDB northbound past Black River Junction Tower 8-22-57. It is running light to pick up an eastbound train./Photo by John C. Illman.

*Two Freight Motors on the Deer Lodge ready track. E42 ACB and E36 ACB in the maroon, orange and black paint scheme, 6-14-63./*Photo by R.V. Nixon.

*Freight Motor E42 AB changing crews at Alberton, 8-12-61./*Photo by R.V. Nixon.

*Extra E42 west near Alberton, 8-12-61./*Photo by R.V. Nixon.

Two units of diesel #80 are shown just after completion at the EMD plant in LaGrange, Illinois. This engine was used in a diesel versus electric economic study on the Coast Division in 1948. Her primary use was branchline log train service./Photo courtesy EMD.

Convinced that electrification was not an obsolete technology, Laurence Wylie immediately began working to save electric operations. His first move was to order a comparative economic study of the electrification versus diesels. It was the first such study made, since Beeuwkes had always accepted the arguments put forward by the Mechanical Department and the analysis from EMD.

J.P. Kiley was startled by the results of the study. Tests using F3 diesel #80 in mainline freight service revealed that, if 3-unit diesels had been used to pull the Coast Division freight traffic, the cost would have gone up. In 1946, 47, and 48, the average annual costs would have been $104,000 higher than what had been spent running the old Freight Motors, 1915 style substations, and maintaining trolley wires. If motive power for passenger trains, ski trains and work trains had been included in the comparisons, diesels would have come out even worse. Costs on the Rocky Mountain Division could be expected to favor electrics by a greater amount. The major factor which drove up the cost of diesel operation was the cost of purchasing diesels. The electrics were, of course, long paid for. Kiley backed away from the idea of scrapping the electrification. However, he would not commit to investing in its modernization without further study and proof of its cost effectiveness. Modernization would at the very least entail the purchase of new electric locomotives.

Laurence Wylie's original move to purchase the 20 surplus Little Joe electrics GE built for Russia was vetoed by Kiley. He was later allowed to purchase 12, but Wylie's plan to have GE build additional new ones was never approved. This brought about a need to find other ways to modernize the locomotive fleet. Wylie gave a hard look at the existing locomotives and what their potential might be if rebuilt. Spurring him on in this study were the continuing efforts by the Mechanical Department and EMD to have the electrification dieselized.

In 1949, a four-unit F7 diesel demonstrator was tested on the Milwaukee. It arrived in Tacoma with a train of record tonnage from Othello, and the salesman talked it up. An article even appeared in a Seattle newspaper stating that modern diesels like the ones pictured in the newspaper would soon replace the obsolete electrification.

Laurence Wylie, whose office was in Seattle, thought otherwise. The type of comparisons being made by EMD were flawed. Not only were new diesels being compared with 30 year old electric locomotives, the diesels were four-unit sets while the electrics ran in threes. Wylie felt that if the Milwaukee were running modern electrics, diesels would simply be unable to match them in any way. What is more, if the existing Freight Motors were run in four-unit sets, EMD would not want to talk about the results of a comparison.

E 64 AB 6-7-46./Photo by Albert Farrow.

E29 at Deer Lodge, 3-73. Note differences between this Freight Motor and E64 as it appeared 27 years earlier./Photo by Noel T. Holley.

E25A at Tacoma Yard 9-3-62. She has been equipped with diesel MU hoses and sockets. She also has gutters on the sides to cope with Coast Division rain./Photo by Doug Cummings.

E40A at Cedar Falls 9-1-62. Her front "porch" has not been expanded for side exit./Photo by Doug Cummings.

E50B at Tacoma Yard 9-3-62. She once ran as a single unit Freight Motor. The mounts for her second pantograph can be still seen on the roof above the road number. /Photo by Doug Cummings.

Middle Unit E45D. At the rear she is coupled to the next unit with a drawbar. At the front, she uses a coupler. On the roof are mounts for a second pantograph. /Photo by Doug Cummings.

Middle Unit E47C at Beverly 4-9-66. She has extended roof walks for the benefit of crews who must jump between units. Although she did not run as a lead unit, she has a snowplow pilot and a coupler./Photo by Doug Cummings.

Middle Unit E47C as seen from the engineers side at Tacoma Yard 4-64./Photo by Doug Cummings.

Middle Unit E25D at Othello 4-9-66. She previously ran as E28C, 10505C, 10505B, 10510C, 10505A and 10210A./Photo by Doug Cummings.

E27C at Deer Lodge 5-29-64. Her windows are half welded over and her side door has been removed./Photo by Doug Cummings.

E34D, a somewhat typical shortened middle unit. Butte, 5-27-64./Photo by Doug Cummings.

E33C carried a pantograph for part of its career after it was shortened. The pan mounts are still evident. No other short middle units carried pans. Tacoma 4-64./Photo by Doug Cummings.

CLASS EF-5

These were the four-unit Freight Motors. They were assembled as part of Laurence Wylie's low cost stop-gap modernization program of the early 1950's. They were intended to help hold the line against the onslaught of diesels until new electric locomotives could be purchased.

They exceeded a million pounds in weight and in terms of tractive effort they matched up very closely with four-unit sets of F7's or GP9's. In terms of horsepower they exceeded the diesels. EF-5's produced 6,680 hp continuously and their short time rating was 8,200 hp. In terms of horsepower measured at the rail, four F7's produced 5,100 and four GP9's produced 5,950.

The true measure of a locomotive's mettle is its tonnage rating and this is where the EF-5's really shined. On a 1.7% grade, it took six F7's to pull 3,330 tons or five GP9's to pull 3,325 tons. A four-unit Freight Motor exceeded all of them with a rating of 3,500 tons. The same was true on a .4% grade where EF-5's were rated at 12,000 tons while six F7's rated 11,640 and five GP9's rated 11,650.

The four-unit Freight Motors were designated EF-5 because ten of the newly purchased Little Joes had already been designated EF-4's. EF-5 was a broad classification covering those motors in which all middle units had cabs, those in which no middle units had cabs, and those in which one did and one did not. As with the earlier three-unit Freight Motors, A's and B's remained the control units on each end of a locomotive. The middle units were C's and D's. The only exceptions were sets E22 ABCD and E32 DACB.

The modernization program for Freight Motors included many ingenious elements. The traction motor armatures were banded with high strength steel wire to protect them against centrifugal force, and traction motor shunts were increased from one to three. This allowed the old Freight Motors to safely achieve speeds of 45 mph. A voltage increase of roughly 10% in the trolley system provided the extra current needed to speed them up.

Other improvements included the installation of fast acting JR breakers to protect the electrical circuits, solid gears to replace the high maintenance spring gears, permanent snowplows in place of the boiler tube pilots, and three to four low maintenance sealed beam bulbs in each headlight. The new maroon and orange paint scheme was also applied to the Freight Motors at this time.

E25 was the first of the Freight Motors to be upgraded and it was also the first EF-5. This set, which was arranged ADCB, rolled out of the Tacoma shops in January of 1951. E25 had previously operated as an EF-3. It became an EF-5 when E28C was renumbered and added to the set. Four more EF-5's followed in 1951, one in 1953, five in 1960 and two in 1961. The only EF-5's to come into existence after 1961 were a few in which not all units bore the same road number; locomotives such as E42 AC/E49 CB. These combinations were created as individual units wore out and those which remained were coupled together for continued operation.

The speed increase helped the Freight Motors to compete with diesels on a more or less equal basis across the Coast Division. Grades and track curvature on the coast made high speed running impractical. Thus Freight Motors with a 45 mph speed limit could match the over-the-road times made by potentially faster GP9's. The GP9's were capable of 65 mph. On the Rocky Mountain Division things were different. The Freight Motors were regarded as slow even at 45 mph. In Montana, there were long sections of track where Little Joes routinely hauled freight at 50-70 mph. On this division the speed increase was useful, but it did not put Freight Motors on the head ends of "hotshot" freights.

The addition of JR breakers was a big improvement, recalls retired engineer Andrew Rogerson.

> "Any of the Motors would flash over the traction motors if you put too heavy a load on them. We used to have problems with that until someone came up with something called a JR breaker. It would kind of meter the load and kick out if you overloaded the motors. It saved a lot of money they would have spent rewinding traction motors."

The idea of running Freight Motors in four-unit sets was discussed while Rainier Beeuwkes headed the electrification according to Walter Gordon.

> "Beeuwkes was against running four-unit Boxcabs," says Walter. "Beeuwkes said the control circuit voltage wasn't high enough to work the controls on four units. After the Milwaukee started running them in fours they did have some problems, but the problems weren't all that bad."

Making Freight Motors run together in fours was nothing compared to the problems in making them run with diesels. During the mid-1950's, more and more diesels began appearing in electrified territory. They were needed to help move increasingly heavy freight trains but their incompatibility with electrics caused problems. Their control systems operated quite differently from those on any of the electrics. Thus they could not be operated in multiple with electrics, and required separate train crews. The Mechanical Department viewed this as an obstacle in the way of efficient operation and proposed a solution. According to Laurence Wylie, their proposal was to have the electrification placed under Mechanical Department control and to replace it with MU compatible diesels.

*E25 ADCB was one of several Coast Division Freight Motors which received "Wylie Throttles" in order to MU with diesels. The others were E22, E39, E47 and E50. Photo date 9-14-63./*Photo by Doug Cummings.

An eastbound freight crosses the Columbia River at Beverly on July 21, 1963. The four unit Freight Motor on the point is assisted by two GP-9s./Photo by Bruce Black.

Pelican E50 pulls drag freight #266 through Kohrs on the way to Deer Lodge 7-17-72./Photo by Ted Benson.

Wylie was the Electrical Engineer when this proposal was made and his response stunned them. First, he successfully argued at the corporate level that diesels in electrified territory were only helpers. As such they rightfully belonged under the control of the Electrical Engineer. Second, he invented an electric/diesel MU throttle. Diesels were then controlled through the use of a control panel and small diesel throttle mounted in the cab of an electric. The diesel throttle was linked to that of the electric with a lever. As the electric throttle was notched out, the diesel throttle was also notched out by a proportional amount. The engineer could control the two types of motive power separately by disconnecting the lever.

These "Wylie Throttles" were first applied to the Little Joes shortly after Wylie retired in 1956. Later they were also applied to several of the EF-5 Freight Motors operating on the Coast Division. These were E22, E25, E39, E47, and E50.

> "The diesels only stayed under Electrification Department jurisdiction for a few years," mused Wylie, "but the throttles stayed until the end."

1963 was the last year that the Freight Motor fleet remained basically intact. Seventy-six of the original eighty-four GE units were still in operation. They were operating in the form of 22 locomotives running on both divisions.

March of 1969 at Avery, Idaho. Motive power from 1915, 1948 and the 1960's is well represented./Photo by Dick Dorn.

Extra E32 west leaving Othello 7-1-59./Photo by R.V. Nixon.

E50A, E50B and E47C make their last run to Deer Lodge. Due to metal fatigue, part of the frame of E50A has broken off. It is still attached to E47C by the drawbar. Alberton, 6-21-73. /Photo by R.V. Nixon.

E25 BCDA and two diesels run in multiple-unit through the arid semi-desert lands east of Beverly. This MU capability gave the aging Freight Motors more flexibility. Photo date 9-14-63./Photo by Doug Cummings.

Under leadened skies, the Butte Helper rests. She has battled blowing snows to help a freight across the Rocky Mountain grade. She is E36A/E37C/E37D/E36B. The date is April 27, 1958./Photo by Doug Cummings.

The 214 mile Coast Division was served by seven four-unit sets and five three-unit sets. Generally ten sets operated as road power and were assisted by a few GP9 diesels. Two sets operated as mountain helpers. The helpers were based in Cedar Falls, at the foot of the Cascades, and in Beverly, at the foot of the Saddle Mountains.

The 440 mile Rocky Mountain Division had four four-unit Freight Motors, five three-unit motors and two single unit motors. This group served along with the twelve Little Joes and several GP9's which were used with the Joes. The single unit Freight Motors were in use as Harlowton Switchers. Generally four of the Freight Motors were used in helper service, two based in Avery to assist in the Bitteroot Mountain crossing and two based in Butte to assist in crossing the Rockies. The remaining five were road locomotives.

Relative to Little Joes, the Freight Motors were slow moving power thus they tended to be used in types of service where speed was not important. They hauled "dead freights," the slow trains which made set outs and pickups along the line, extra freights, and of course, served as helpers. The Freight Motors were extremely powerful in the mountain climbing speed ranges of 12-14 mph. At those speeds, a four-unit set could handle more tonnage than two Little Joes. Additionally, they could do it while drawing only as much current as one Little Joe. Comparatively, the strengths of the Little Joes lay in hauling tonnage at speed. They could easily roll trains at speeds in the 50-70 mph range which was unattainable for the Freight Motors. Originally, the Freight Motors had been upgraded under the assumption that they would only need to last a few more years. Although they were worn and obsolete, the cost to upgrade them was low. Operating costs were, however, climbing. The Freight Motors needed more and more parts replaced in order to continue in service. Since the Milwaukee had to make its own parts the cost was high. Beginning in 1951, any units severely damaged or excessively worn were retired, cannibalized for parts and then scrapped.

The first GE units to go were E51 AB and E68 AB in January of 1951. A wreck in the Saddle Mountains scratched E51 from the roster, while in the case of E68 it was a head-on collision with a Union Pacific 2-8-2 near Auburn, Washington. In 1953 E41 ACB was involved in a runaway at Doris. This brought the retirement of the badly damaged C and B units.

From then through 1963, all units continued to roll, but no new electrics arrived to replace them. Year after year the efforts by the Electrification Department to purchase new locomotives were turned down. By 1963, the crisis stage had been reached and wholesale scrapping began. In 1964, thirteen units were retired. It was sixteen in 1965 and several each year thereafter. By 1970, only 31 of the original 84 units were still running and these were wearing out fast. Everything was going out at once. The superstructures of some Freight Motors were rusted through in spots, the frames suffered from stress and metal fatigue and the electrical equipment was plagued by short circuits. Among the employees, respect for these once great locomotives dwindled.

> "Metal fatigue was breaking the frames," recalls retired Carpenters Helper, Wade Stevenson. "One time at Othello, a guy coupled onto his train real hard and the frame of his Freight Motor just broke into pieces and fell to the track. Those frames were welded and rewelded many times."

At places like East Portal, in the Bitterroots and Hyak, in the Cascades, substation operators complained that the Freight Motors were responsible for a lot of electrical faults on the line and they often tripped out the substation circuit breakers.

E50A/47C/35C/50B prepares to leave Avery for Deer Lodge and a visit to the shops. Her train was so heavy that the crew wanted a helper. When the dispatcher refused to assign one, E50 rolled into the night alone, her train moving barely faster than a walk. Summer 1971./Photo by Noel T. Holley.

A westbound dead freight rolls by Finlen, Montana on June 30, 1970. It is pulled by E47 BC, an unusual Freight Motor set. It is unusual because E47C has no cab controls and no pantograph./Photo by Bruce Black.

E39 ADCB and a westbound freight await a meet at Hyak Yard on August 2, 1966. Ahead lies the 22 mile descent through Snoqualmie Pass./Photo by Doug Cummings.

Yard goat E57B goes to work on the PM shift at Harlowton, Montana. Engineer is draining his air reservoirs (just like steam days!). May, 1974./Photo by Ted Benson.

In Tacoma, the weekend shop foreman had little regard for Freight Motors. He proudly offered to show me one of his new GP40's but when I asked to see a "Pelican" he mumbled and muttered.

"Junk," he growled. "They told me they were going to get rid of this trolley stuff a long time ago but they haven't done it. Are you sure you wouldn't like to see a GP40?"

When I asked again about a "Pelican" he corrected me while leading me to the building where E25 was being prepared for an after midnight run.

"We call them GE's" he said. "Never heard anyone call them Pelicans. They're worn out and they're slow but I'll give them one thing, they can pull anything you can put behind them. You can't overload them, they just run slower. If you overload a diesel you'll burn up the traction motors."

On that rainy spring day in 1970 E25 sat with its nose sticking out of a long brick shop building. Its maroon and orange paint was badly faded and the running gear was caked with grayish-brown dirt. To me, it looked good, but in two short years it would be hauled to Deer Lodge, retired and stripped for parts.

The end came quietly on the Coast Division. Electrics were only occasionally operated there after 1968, and by 1970 only three sets were left. E50A/E47C/E35C/E50B was sent to operate in the Rocky Mountain Division in 1971. During early 1972, E25ADB was sent to Deer Lodge and scrapped. E39A/E39D/E39C/E47A was retired at the end of the year. With the passing of E39 the Coast Division electrification was dead. Time and diesels had won.

On the Rocky Mountain Division the Freight Motors did not go down without a fight but

*Six units muscle a westbound "dead freight" over the Rocky Mountain summit at Donald. According to Wade Stevenson, the Milwaukee drew up plans for a six-unit motor but never built one. This set is helper E34 ACDB and road locomotive E47 BC. The date is June 30, 1970./*Photo by Bruce Black.

A view from the rear cab of E45 as it works in mid-train helper service. It is helping E70 and E75 westbound across the Bitterroots with 5,100 tons. Summer 1972./Photo by Noel T. Holley.

At a point just south of Seattle, the Pacific Coast RR tracks the Milwaukee ran on were sandwiched between N.P.-G.N. line in the foreground and the U.P. in the background. E47A/E39DCA is running northbound on April 7, 1971./Photo by Jack Powers.

keeping them running was hard. In 1972, the mood of Deer Lodge shop foreman Ralph Hagemo was optimistic but worried:

> "They told me that they are going to have some new motors for me if I can just keep these things running a few more years," he said. "If the electrics go, this shop will be closed and we'll all be out of work. As long as I have anything to do with it these things are going to run until they can't turn another wheel."

In January of 1972, only five well worn Freight Motors remained in service on the Rocky Mountain Division. Four of them were assigned to helper service and occasionally to dead freights or work trains. These were E29A/E36C/E29B, E34 ACB, E45 ACB and E50A/E47C/E50B. The remaining locomotive E57B/E47D worked as the Harlowton Switcher. In December, E47D acquired a large dent in the rear of its cab when it accidentally rammed some freightcars. Officials ordered it scrapped, but since it was not worn out, Hagemo was reluctant to carry out the order. As a result, the worn out E34C was sacrificed to the torch minus numbers. E47D was repaired and returned to service bearing the number E34C.

In October of 1973, E34 A & B were retired due to wear. That month also saw the end of E45 A & B. E45C had been retired two months before. E50A/E47C/E50B was retired in August. It was a tide that could not be stemmed by the hopes and will of the Deer Lodge shop foreman. One by one, the locomotives which had once been the world's mightiest turned their last wheel.

Avery Helper E45 ACB growls eastbound through East Portal and prepares to go into regenerative braking. August 1973./Photo by Noel T. Holley.

In 1973, E29A/E36C/E29B was the only Freight Motor in road service. One day in April while pulling a train its frame broke. The set was unceremoniously removed from the roster.

On June 15, 1974, when the end of electric operations came, only E57B/E34C were still running. Serving in the undemanding role of switchers, they outlasted their road service sisters. Each had served for 58 years.

The Milwaukee preserved three units from this once great fleet for posterity. E50 A & B are in Duluth, Minnesota at the Lake Superior Transportation Museum. They were the first of the Freight Motors and their present appearance is very similar to the way they looked upon delivery from General Electric. They are painted black and renumbered 10200 A & B. The third remaining unit is E57B. It sits in downtown Harlowton and was painted maroon and orange when sent there. E57B was built as 10211B.

The GE Motors were a rugged and for many years a trouble free fleet. When Laurence Wylie looked back upon them in 1975, he called them the best electrics the Milwaukee ever owned.

10200 (E50) at the Lake Superior Transportation Museum in Duluth, Minnesota./Photo courtesy Milwaukee Road.

A mile or so of Lewistown Branch was under wire at Harlowton where yard motor E57B/E34C offers an idea of what it might have looked like had the Milwaukee decided to electrify a branchline. By May, 1974, with E57 the last boxcab operating anywhere, such esoteric considerations were fine for dreamers and no one else. Most locals were more concerned with a display for the E57 once the Milwaukee threw the circuit breaker forever. /Photo by Ted Benson.

Abstract view of E49B outside the shop at Tacoma's Tideflat Yard in July, 1972. By this time, E49 was good for parts and at that, there wouldn't be much demand for boxcab components after November. /Photo by Ted Benson.

End of the Line. The Deer Lodge scrap line in September 1973./Photo by Noel T. Holley.

Over by the shop, E25A rotted in total disgrace. I remembered her proud pose for Stein's camera at Othello for Western Trains...tonight she hid far from public scrutiny, cast out like some broken Lionel toy. She deserved better. Deer Lodge offered no salvation. 5-74./Photo by Ted Benson.

CLASS EP-1A PASSENGER MOTORS E22 AND E23

This class consisted of two locomotives. They were ex-Freight Motors which were rebuilt and regeared for passenger service during the 1950's.

In 1952, the Milwaukee's electric passenger locomotive fleet consisted of Bipolars, Westinghouse Motors and Little Joes. The Coast Division was served by five Bipolars. The Rocky Mountain Division was served by two Little Joes and the seven remaining Westinghouse Motors. Except for the Little Joes all were well worn and technologically obsolete. Due to the discontinuance in 1947 of trains #7 and #8, the Spokane to Butte Local, not all of these motors were needed.

Laurence Wylie wanted to replace all of the old freight and passenger locomotives with Little Joes. By 1952 it has become clear to Wylie that the purchase of more Joes was not imminent. Thus he would have to make do with the locomotives he had. Although Wylie felt that the Bipolars could be modernized and retained in service, he did not feel that way about the Westinghouse Motors. They were fast and well-liked by train crews but they had always been troublesome to maintain. Little Joes E20 and E21 functioned quite well in passenger service, but it was felt that they could be utilized better as freight locomotives.

In order to facilitate removing the Joes from passenger service while phasing out the Westinghouse Motors, some other type of passenger locomotive would have to replace the Joes. Wylie's solution was to resurrect the EP-1 class GE Passenger Motors. From 1915 to 1920 when the original Passenger Motors were in service, they acquired a reputation for reliability and good operation. Wylie felt that if they were returned to service with some modern improvements they would be even better than before. The first locomotive selected for conversion was E69 A & B. It had originally been contructed as Passenger Motor 10103 A & B. Since 1920, it had worked in freight service bearing several different road numbers.

E69 A & B was hauled into the Tacoma shops for modification and in February of 1953 E22 A & B emerged. The most obvious change was the addition of a rounded front on E22A. There were, however, other changes. Added were solid gears with a 2.4 to 1 ratio, roller bearing pilot trucks which came from scrapped F-7 class 4-6-4's, air brake systems of the same type used on Little Joes, JR breakers in the electrical circuits, lightning arrestors on the roofs, shock absorbers under the motor-generator sets, and Clarkson steam generators. These steam generators were the same type used in diesels and they are often called "flash boilers." The locomotives also contained enough fuel and water to heat a train for up to eight hours. At 326 tons, E22 A & B weighed nine tons more than the original Passenger Motors. The official horsepower rating was also higher. The EP-1A rating was 3,700 hp continuously and 4,800 hp for one hour.* The original EP-1 class was rated at 3,000 hp continuously and 3,440 hp for one hour. E23 A & B rolled out of the Tacoma shops in August of 1953. It was identical to the E22 set except that it weighed six tons more. It was converted from Freight Motor E28 A & B.

E22 and E23 were a source of great pride when they were first rebuilt. The cosmetic metal work and glossy new paint applied by the shop crews gave them a modern slightly European appearance. Wylie's projections on theoretical horsepower also contributed to the appearance of modernity. There was, however, a problem. Their performance in service did not measure up. Not even Wylie batted 100% in making silk purses from sows ears and in later years he had little to say about E22 and E23.

*Horsepower rating was later officially revised downward.

Passenger Motor E23 AB between trains at Othello. It is wearing Union Pacific Streamliner colors and EMD side grills. July 1, 1959./Photo by R.V. Nixon.

Under typically leaden Seattle skies, E23 AB prepares to pull out of Union Station with an eastbound passenger train in 1958./Photo by Otto Perry.

E22 DCBA roll east through Kittitas on September 2, 1962. The two units on the point were previously known as E23 AB./Photo by Doug Cummings.

Milwaukee #16 heads eastward through the interlocking at Black River behind E23. 8-22-57./Photo by John C. Illman.

Barry Kirk was one of Wylie's assistants in 1953 and later headed the Electrification from 1963 to 1971. As Kirk tells it:

"E22 and E23 were not a big success. They couldn't make the necessary speed. They were rated at 70 mph on level track but they couldn't exceed 35 mph on a 1% grade. They didn't compare with a Westinghouse Motor or a Bipolar for use on passenger trains. Crews started complaining about that almost as soon as the Passenger Motors went into service, but Wylie just passed it off as idle grumbling and refused to believe it. Crews kept on complaining and so one day I decided to ride the locomotive to Othello and see what was happening for myself. When I got back, I told Wylie it wouldn't make speed but he didn't want to believe me either. He got mad but he did agree to ride in the cab and see for himself. After we arrived in Othello he still couldn't believe it and he said the shop crews must not have rewired it according to his plans. He kept me up all night going through that motor circuit by circuit with him. In the end he had to admit that it was wired right but it just wouldn't pull. A couple of years later we added a third unit to E23 in order to give it more power."

E36C, a middle unit with a cab, was rebuilt in 1955 to serve as part of E23. As E23C, it was the only middle unit to operate in passenger service.

The Olympian Hiawatha, running as Pacific Coast #27, is hauled backwards between Black River and Seattle by E23. 8-22-57./Photo by John C. Illman.

The most often heard complaint about E22 and E23 concerned the rough ride they gave their crews. Pat Chester rode them as a Coast Division fireman and he has choice words about them:

> "They didn't ride worth beans and they were some of the most miserable things in the world to ride on. A fireman would have to walk back and patrol the locomotive at regular intervals during the trip. At 60 mph, your elbows and ribs would be beat to a pulp when you came back out of those narrow aisleways. The Milwaukee geared them up to run fast, but it was like putting a V-8 engine in a Model T. It didn't make them modern. Crews didn't like them, but in all fairness they just weren't a Bipolar or a Fairbanks-Morse, they were a bunch of worn out junk."

A service change which came about in 1955 was the discontinuance of local mail trains #17 and #18, the Columbian. With their demise there was only one passenger train per day each way in electrified territory, the Olympian Hiawatha. It was generally pulled by Little Joes, Bipolars and GE Passenger Motors.

In 1958, Little Joes E20 and E21 were assigned to freight service. Train crews struggled with unreliable Bipolars and rough riding GE Passenger Motors until 1959 when diesels were assigned to the train. The electrics were then held in reserve until the Olympian was discontinued in 1961.

After the end of passenger service the GE Passenger Motors were converted to freight. E23 A & B were coupled behind E22 A & B to create a four-unit Freight Motor. With round nosed units at each end, it was unique. Renumbered E22 ABCD, it was the only multi-unit GE Motor which had its units in alphabetical order.

According to Wade Stevenson, a retired Othello shop employee, E22 retained its high speed gearing for a while after entering Freight service. It was regeared for freight after showing an inability to start high tonnage trains.

> "In freight service the crews would try to run them just like the other Freight Motors," says Pat Chester. "When they did that on the high geared E22, they were either burning up the traction motors from overloading them, or snapping trains in two from starting too fast."

There was another problem with Freight Motor E22 recalls Chester:

> "Since there were no doors in the ends, it was hard to get back to the diesels and patrol them while the train was moving. If a diesel was running low on oil or water and overheating, the fireman had to go back there and work on them. To do that you had to climb out the side door and walk along a narrow ledge to a point where you could jump across to the diesel. With my short legs I could just barely make it."

E22 operated in Coast Division freight service as a four-unit set until 1966 when E22C was wrecked. The three remaining continued to operate until 1968 when they were retired and all four in the set were stripped for parts. They were finally scrapped in 1973. The unit which had operated in passenger service as E23C was renumbered E39D in 1961. It operated until 1972 when the E39 set was retired.

E22 BA at th new Tacom Station o August 18 1958. She wi pull the trai backwards t Seattle's Unio Station. Ther she will move t the head en and pull it for ward on the tri east to Othello 8-18-58./Phot by Wally Swan son.

E23C at Tacom Yard 12-27-5 Note she equipped with headlight, horn bell and snov plow pilot fo lead unit opera tion. She als has an old pil truck with le springs rath than a ne one with co springs. The s was original intended operate in ABC arrang ment, but spe most of i career arrang ACB./Photo Doug Cun mings.

E23B at Tacor Yard, 12-27-5 /Photo by Do Cummings.

E22A at Tacoma Yard 3-31-62. She is MUed with a GP-9./ Photo by Doug Cummings.

'2B at Cle um 5-20-67./ ıoto by Doug ımmings.

'2D at Cle um 5-20-67./ ıoto by Doug ımmings.

E39D at Othello 7-1-66. She is the old E23C with a new number./Photo by Doug Cummings.

Time was just about up for the box motors left at Tacoma in July, 1972. E39 would claim the honors of being the last to operate on the Coast Division in a matter of weeks—for E22B, the race is already run./Photo by Ted Benson.

CLASSES EF-1 AND ES-3 THE SINGLE UNIT FREIGHT MOTORS

The GE Motors were designed to allow the independent operation of single units and sometimes the Milwaukee ran them that way. A coupler was mounted in the rear drawbar pocket, footboards were installed front and rear, and a headlight was mounted on the rear roof. If the locomotive was to be used in road service a second pantograph was mounted above the clerestory, to provide a back-up in case the pantograph in use was damaged.

Because the Milwaukee served so few towns and lineside industries along the electrified main, there was little need to use single units in local service. They were most commonly used as work train motive power and for yard switching. As switchers they were used almost exclusively in Harlowton. There, switchers sometimes needed the ability to move entire trains through the yards. As train weights grew, the Steeplecab ES-2's were unable to do the job, but the larger, heavier, single unit Freight Motors could. By the end of 1951, the Milwaukee stopped using Steeplecabs in Harlowton. From then on the Harlowton Switchers were always Freight Motors.

Bill Lintz, a retired engineer from Deer Lodge recalls the difference in power between a single unit Freight Motor and a Steeplecab.

> "The single unit Freight Motors used to come to Deer Lodge for maintenance and repairs. We usually tested them here before sending them back to Harlowton and so sometimes I used them for switching. The Deer Lodge yard was built on a slight grade and going into the yard lead a Freight Motor could handle 6,000 tons. A Steeplecab could only push 4,500."

In terms of 50 ton freight cars, that comparison meant 120 cars compared to only 90.

During most of the time the electrification existed, the Milwaukee classified all single unit Freight Motors as EF-1's. In the 1960's however, when the only single units in use were Harlowton switchers, the Milwaukee began designating them class ES-3. This designation continued in use even after 1970 when cabless unit E47D was teamed with E57B to form a two unit switcher. E47D was damaged in a collision in 1972. When repaired it was renumbered E34C. The duo of E57B and E34C remained in service until 1974 when electric operation was phased out.

A partial list of those units which served singly is as follows:

> 10201A, 10201B, 10202A, 10202B, 10203A, 10208A, 10226B, 10236A, 10236B, 10240A, E55A, E56B and E57B.

The unaccounted for unit in these split pairs was usually operating as a C or D unit in a locomotive with a different road number. No more than four units operated singly at any one time.

The Harlowton Switcher, E57D/E47D pose for a portrait on October 4, 1970./Photo by Bruce Black.

THE STEEPLECAB SWITCHERS
CLASSES ES-1 AND ES-2

The Milwaukee operated only five electric locomotives which were designed and built as switchers. They were the ES-1 and ES-2 class steeplecabs, a group of B-B locomotives which resembled the freight motors GE built for interurban lines. With their center cabs and sloping hoods they provided excellent visibility in all directions for switching crews.

ES-1, THE GREAT FALLS STEEPLECAB

The first of the switchers was the one and only ES-1. It was built in 1915 as the 10000 and was sent to work the streets of Great Falls, Montana, 195 track miles from the mainline at Harlowton. The 10000 was needed in Great Falls because of a 1912 street railway ordinance banning steam engines from the streets of town. Falls Yard was outside the city limits, so the steam engines switching there were not affected. They could not, however, serve the industries along Valeria Way. To solve this problem, the Milwaukee spent $50,000 electrifying its in-city trackage. Within Great Falls, four miles of branchline and three miles of spurs and sidings were electrified at 1,500 volts DC. The Steeplecab locomotive 10000 weighed 50 tons, produced 300 horsepower, and could pull nine to ten loaded freightcars at 9.5 mph. It was powered by four 750 volt traction motors wired in pairs for 1,500 volt operation; thus it was not readily adaptable to the 3,000 volt mainline system. It rode on solid steel wheels 36 inches in diameter and the locomotive body was somewhat lower than those of the later ES-2's. The pantograph

Switcher Ten-Thousand was the only locomotive in the ES-1 class. She is shown at the edge of town in Great Falls. The year is approximately 1915./Photo courtesy Milwaukee Public Library.

used was a one-of-a-kind type on the Milwaukee with cross folding legs. It was mounted on a tall platform above the cab roof. Because the locomotive body was low, a platform was needed in order for the pantograph to reach the Milwaukee's 24 foot high trolley wire.

Electric switching in Great Falls continued for 22 years, then came to an abrupt halt. In July of 1937, the Great Falls city council amended its street railway ordinance to allow the use of oil burning locomotives in town. The act was the death knell for the 10000. The Milwaukee had long viewed its existence as an unnecessary expense. It assigned an oil burning steam engine to intown switching and scrapped the tiny electrification. By September, the wires were down and ES-1 10000 was for sale. Though it was offered at scrap metal prices, no buyer was found on the used locomotive market. In November, 1939, it was scrapped. Eight months prior to scrapping, the 10000 was assigned the number E85 as part of the system-wide renumbering program. Since the locomotive was out of service at the time, that number was probably never painted on it.

ES-2, THE MAINLINE STEEPLECABS

The first two ES-2 types, 10050 and 10051, were built in the summer of 1917. They were 3,000 volt switchers intended for use in Butte, Montana. Butte had a small yard and a number of industrial sidings. These two locomotives weighed 70 tons each, were rated at 500 horsepower, rode on 40 inch spoked drivers and had a top speed of 35 mph.

As a result of the proven usefulness of the Butte switchers, the Milwaukee purchased two more ES-2's in 1919. They were part of the order which brought the Bipolars and half of the Coast Division substation equipment. These last two switchers were numbered 10052 and 10053. They were identical to the first two except that they carried additional steel plates which raised their weight to 82 tons. This gave them more tractive effort. They worked on the Coast Division initially. One was assigned to Othello and the other one possibly to Cle Elum. Neither stayed on the Coast Division very long. Retired engineer, Bill Plybon, ran one in Othello around 1920. As he remembers it, the trains quickly became too heavy for the Othello switcher and it was sent east to the Rocky Mountain Division. The four ES-2 locomotives were designed to operate singly and primarily at low speeds, they were not wired for multiple unit operation or regenerative braking. For convenience in maintenance, the pantographs, headlights, cab heaters, and small switches were of the same type used on the GE Motors. Another feature, which was the same as the GE Motors, was the "fish pole" or "stinger." This was a trolley pole used for operating the air pump and motor-generator when no air pressure was available to raise the pantograph. Since the fish pole was never used when the locomotive was in motion, it carried a flat bar at its tip rather than a wheel or a slider shoe.

When the ES-2's were first built the fish pole was mounted on one side of the pantograph frame. It was later moved to the cab roof and mounted on insulators. When the pole was down, it was grounded by a long hold-down hook extending up from one locomotive hood. The hook contained a locking device to prevent the pole from bouncing loose and raising accidentally. Whenever the crews raised the pole it pressed against a springy sheet metal plate which connected it with a 3,000 volt conduit on the locomotive roof.

Mounted on the locomotive's other hood was a horn-gap lightning arrestor. It sat level with the cab roof and was supported by long metal legs. Its purpose was to allow high voltage such as lightning to reach ground by arcing across a gap between two wire horns. Because of this

Brand new switcher Ten-O-Fifty-Two posed for an official General Electric photo in 1919. Note that she was built with her "fish pole" mounted on the fireman's side of the pan frame. She also did not have an external lightning arrestor. This unit later became E82./Photo courtesy Milwaukee Wisconsin Public Library.

E80 switches passenger cars on a February day at Butte in 1940./Photo by Dan Dougherty, Casey Adams Collection.

quick route to ground, lightning was less likely to ground itself by arcing through the traction motors and control equipment. During the 1950's four sealed-beam headlight bulbs were mounted in each headlight. Because the level of brightness provided by all of these bulbs exceeded what was needed for switching, the headlights were often operated with only two bulbs.

The numbers assigned to these locomotives remained the same until March of 1939 when as a result of the Milwaukee's system-wide renumbering the ES-2's were renumbered E80 through E83.

The four-locomotive ES-2 fleet remained intact until 1952 when E83 was scrapped. It was declared surplus and cut up for parts, according to retired Rocky Mountain Division Master Mechanic, Dean Radabaugh:

> "We had four in service and we didn't need the fourth one anymore. The Milwaukee had been using one in Harlowton but it wasn't heavy enough for that service. It was replaced with a single unit Freight Motor. We only needed one in Butte, one in Deer Lodge and one held out for maintenance, so the fourth one was scrapped."

E80 through E82 continued to operate until the end in 1974. Since they were not worn out, they were not replaced until the wires came down.

According to retired Chief Electrical Engineer, Laurence Wylie:

> "The switchers were the simplest motor of all. They were operated for many years at a low maintenance cost. About the only real change we made in them was to add steel plates to make them heavier and stop the wheel slippage."

Barry Kirk, one of Wylie's successors recalls that the switchers were a well liked group of locomotives:

E82 in Butte August 5, 1956./Photo by Doug Cummings.

"For a while, when the diesels were new, the crews preferred diesels but after the novelty wore off they asked for us to give them the electrics back. Crews liked the electrics because they could start moving so quickly and they required so little maintenance. In the Butte area we even put in additional trolley so that more industrial sidings could be served by electrics."

Quick throttle response was very important to switching crews who sometimes had to "kick" cars into sidings. The response of electric switchers was almost instantaneous. Notching out the throttle immediately allowed the ever-present power in the trolley to flow through the traction motors and move the locomotive. By contrast, a diesel switcher's throbbing prime mover first has to rev up to a higher speed and generate more electricity before the traction motors can receive more power. This works like a built-in time delay on throttle response.

Yard goat E81 at Deer Lodge July 6, 1973./Photo by Bruce Black.

SWITCHING IN THE ELECTRIC ZONES

Four ES-2 switchers are not very many on a mainline that stretches from mid-Montana to Puget Sound. When all wired sidings are included in the count, the Milwaukee had almost 900 miles of trolley wire above its tracks in Montana, Idaho, and Washington. The Northern Pacific had approximately 80 switch engines serving its line in the same geographical area. There were, however, significant differences between the Northern Pacific and the Milwaukee. The Northern Pacific line, which was built first, passed through all of the major cities, blanketed the logging and mining areas with spurs and served most of the existing industries. Because Milwaukee could not gain easy access to the towns and industries, it passed them by. It generally passed on the outskirts of towns and it had few spurs and branches to generate lineside traffic.

Large switching yards tend to be built for consolidating locally generated traffic, for splitting up groups of cars taking different routes and for interchange with other railroads.

The Milwaukee basically had only one route and its competitors, the Northern Pacific, Great Northern, and Union Pacific, ran paralleling routes. They had little need for interchange with the Milwaukee since they too served Butte, Spokane, Seattle, and Tacoma. While the Northern Pacific had twelve major yards with switchers assigned to them west of mid-Montana, the Milwaukee had only six. These were Harlowton, Butte, Deer Lodge, Spokane, Seattle and Tacoma. Avery and Othello were primarily engine terminals and were switched by road crews using road locomotives.

*Engineer Ken Boynton takes a couple of notches on the controller of E82 as he works the day yard job at Deer Lodge in May, 1974./*Photo by Ted Benson.

Step aside sweetie and let Grannie take the stage...Deer Lodge switcher E80 assumes the mainline in front of the power of #262 to make a brief switching move with the eastbound hotshot. Three GP-40's and a Joe were standard power on 261/262 in July, 1972 and it looked like the steeple cab switchers would last forever. In terms of the electrification, they did!/Photo by Ted Benson.

All six yards with assigned switchers were small compared to those on the Northern Pacific and only three of the six had electric switchers reguarly assigned to them. Harlowton was generally switched by one or two single unit Freight Motors, and for many years an ES-2. Butte and Deer Lodge were served by one ES-2 switcher each.

> ''We needed a switcher in Deer Lodge because it was a classification point,'' says retired engineer Bill Lintz. ''We broke up all of the east and westbound trains there until the later years when the classification work was moved east. After that the only reason we needed a switcher was to serve the Potlatch lumber mill.''

Steam and diesel switchers were used in Spokane, Seattle and Tacoma. Spokane was not on the electrification. Seattle was reached via the tracks of the Pacific Coast Railroad and the Union Pacific. Tiny Van Asselt yard, adjacent to Seattle's Union Pacific yard contained only two wired sidings. Stacy Street yard and the joint use port trackage were not electrified. In Tacoma's Tide Flats yard, the only wired tracks were those needed for attaching and removing road locomotives from trains. Tacoma was the terminal for several branchlines. As a result, there were always a lot of small branchline locomotives sitting around.

As Laurence Wylie pointed out:

> ''There was no need for us to buy electric switchers and electrify the yard when there were plenty of idle branchline locomotives available for switching.''

Steeplecab electric E81 switches the Deer Lodge Yard on the evening of June 13, 1974./Photo by Dick Dorn.

June 15, 1974, the eternal day for Milwaukee electrification finds E82 switching the Deer Lodge Yard. This was the last electric to operate on the Milwaukee./Photo by Dick Dorn.

THE BIPOLARS
CLASS EP-2

General Electric named them the Bipolar Gearless Type, and the name stuck. It was a name used from coast to coast. The poeple who ran them used it and so did the people who had no idea where they ran.

These locomotives were called Bipolar because each traction motor had only two poles. These field poles were mounted directly on the locomotive chassis beside each axle. They were gearless because instead of using high speed motors turning gears on the axles, they had low speed motors with the motor armatures mounted directly on the axles. These features made for a unique and silent locomotive. Since the motor speed was only 458 RPM at 60 miles per hour, there was no traction motor whine, and since there were no gears there was no gear tooth growl either.

Though these were not the first or only locomotives to use the bi-polar drive system, they represented a new height for railroad technology when built. They were briefly the "mightiest electric locomotives in the world," and their size and power attracted a lot of attention in the post World War I years of Model "T's," flappers and noisy smoke belching steam locomotives.

The Milwaukee ordered them in 1917 when the decision was made to proceed with electrification of the Coast Division. Five were built and the series was numbered 10250 through 10254. The first two were completed in 1918 and the others followed in 1919. They quickly achieved a fame that far exceeded their numbers and made them almost larger than life.

Introduced with much show and fanfare, they caught the fancy of America. They came to symbolize the Milwaukee Road and the Olympian, a train called "The Best to the East." They symbolized General Electric and the mainline electrification which lay just around the corner in our future. The Bipolars had a distinct kind of beauty and this coupled with their other attributes made them good for public relations. General Electric hoped that these locomotives would signal the dawn of an electric age for mainline railroads in the United States, a time when all American railroads would replace their steamers with catenary and highly efficient "juice jacks." The Milwaukee hoped that these locomotives, as symbols of a commitment to the newest and the best in service, would attract more business. GE staged locomotive shoving contests for the press and railroad dignitaries at its test track in Erie, Pennsylvania. The Milwaukee staged more on its tracks at Kent, Washington, and high on a curving trestle in the Cascade Mountains. In all cases, the Bipolars easily shoved the struggling steam locomotives offstage. These publicity stunts and a grand tour of the Milwaukee system made the Bipolars a locomotive that the public would remember for years to come. The Bipolars could produce 3,200 horsepower continuously and apply 123,500 pounds of tractive effort when they needed to start a train. They had 12 motors, rode on 28 wheels and at $200,000 each, they cost as much as several steam locomotives.

The Bipolars were designed and assembled by General Electric at its locomotive shop in Erie, Pennsylvania. Construction of superstructures and running gear was subcontracted to the American Locomotive Company. The resultant passenger motors were 76 feet long, 10 feet wide and their height over locked down pantographs was 16 feet 8 inches. The tallest part of these locomotives was, however, the smoke deflector on the train heat boiler. The height over that was 17 feet 3 inches.

The total wheel base was a flexible 67 feet and the rigid wheel base was 13 feet 9 inches. The wheel arrangement was 1B + D + D + B1. The idler wheels on each end of the locomotive were

36 inches in diameter and made of pressed steel. The drivers were 44 inches in diameter and had steel tires fitted on center castings which had eleven spokes each. All four trucks were connected together by swiveling ball and socket drawbars which carried the pulling force between trucks and couplers.

The body of the locomotive was in three sections. The center section was a compartment containing a vertical fire tube boiler, 3,600 gallons of water and 750 gallons of fuel oil to provide steam heat for the passenger trains. This section was not attached to the chassis, but instead supported by brackets on the end sections of the body. The end sections contained nearly identical cabs, in front of which were long, rounded equipment hoods. Inside these hoods was electrical gear arranged around an aisleway which ran the length of the hood. The gear consisted of such things as contactors, control circuits, a lightning arrestor and, in the "A" end, a motor generator set. The M-G set used a 40 horsepower, 3,000 volt DC motor to turn an 80 volt DC generator which provided current for train lighting and low voltage equipment on the locomotive. The Bipolars contained one air compressor and this was carried on the right hand side of the "B" end hood in the rectangular box just ahead of the cab.

The end sections of the Bipolar's body were rigidly bolted to the four axle trucks, below the cabs. Their weight also rested on the three axle trucks, but this was through a set of rollers running on an inclined rack. Besides carrying weight, the rack and rollers acted as a truck centering device. As the locomotive entered a curve, the guiding truck would swing from its spot directly beneath the end of the hood. When it did this, the end of the hood was lifted slightly by its rollers moving up the inclined rack. The lowest point of the rack was the center. When the locomotive left a curve, gravity pulling down on the hood would work to move the guiding truck back to the center point beneath the hood.

An engineer's side view of E3 from the B end at Tacoma Yard, 3-28-39./Photo by Albert Farrow.

The Hamilton Watch Company special at Kent, Washington, 12-14-23. Because of its accuracy, Hamilton was an "approved" brand of watch for railroad men./Photo by Asahel Curtis courtesy Washington State Historical Society.

Bipolar E4 and the Olympian westbound on Houser Street in Renton. Date unknown./Photo by Harold Hill.

Bipolar 10254 westbound approaching Hyak with #15 7-11-25./Photo by Asahel Curtis courtesy Washington State Historical Society.

The Bipolars carried two pantographs, one mounted above each cab. Because the pans were so close together, the rule book directed the engineer to use the rear one rather than the front one. This prevented the loss of both pans if the one in use became entangled with the trolley. The pans were an air raised, spring lowered type. The Bipolars carried an auxiliary battery driven air compressor which could raise a pan if both pans were down and there was insufficient pressure remaining in the air tanks to raise them.

The traction motors on the Bipolar were designed to operate at 1,000 volts each. There were 12 traction motors, each rated at a maximum of 266 horsepower. Amperage drawn by the locomotive was directly related to the tractive effort being produced. For instance, if starting required 118,000 lb. of tractive effort, then the locomotive drew 6,000 amps while starting. During normal running, the locomotive produced 42,000 lb. of tractive effort and drew 888 amps.

At the fastest throttle setting, the 3,000 volt current in the trolley passed through groups of three 1,000 volt motors before it traveled to ground. At the lowest continuous speed the current traveled through each of the twelve traction motors one after another before reaching ground. Each motor received 250 volts.

According to General Electric, the Bipolars were capable of achieving 90 mph. The performance graphs, however, topped out at 65 mph and the Milwaukee Road rule book limited them to 60 mph.

In the regenerative braking mode the Bipolars used ten traction motors for braking power. The two remaining motors were used to generate control current. Bipolars could enter high speed regenerative braking at 22-30 mph and low speed regenerative braking at 11-15 mph.

The Bipolars were designed to pull any normal Milwaukee Road passenger train singly and thus were built without multiple-unit controls. If any train exceeded their capacity, they had to be double headed with a crew in each locomotive. Multiple-unit controls were not added until the Bipolars received a major rebuilding in 1953.

The ski train from Seattle has arrived at the Milwaukee Ski Bowl. The typical snow depth made pedestrian snow sheds a necessity./Photo by Laurence Wylie.

Following delivery from General Electric the Bipolars were briefly tested on the Rocky Mountain Division before they were forwarded to the Coast. When they arrived these silent locomotives with shrill high pitched whistles were an instant success. The 4-6-2's they replaced could only pull 7 cars on the Coast Division's 2.2% westbound ruling grade. The Bipolars proved they could handle as many as 12 to 13 westbound. They could handle 15 to 16 on the 1.7% eastbound grade where 4-6-2's could only handle 9. The trains the Bipolars hauled would have required double heading and extra engine crews if handled by steam. Additionally, Bipolars could run from Tacoma to Othello without stopping to be serviced. Steam engines required several water stops and an engine change in Cle Elum, midway across the division.

Bipolar tonnage ratings varied over the years. They depended on the dispatcher's viewpoint regarding the value of pulling more cars versus making more speed. Regardless of what the assigned tonnage ratings were, the Bipolars continued to best the performance of steam. The Milwaukee's Idaho Division 4-8-4's were rated at 3 to 4 passenger cars less than the Bipolars for Coast Division grades. These 4-8-4's were considerably newer than the the Bipolars and comparable in performance to the 4-8-4's used on other railroads.

From 1919, through part of 1957, the Bipolars operated on the Milwaukee's Coast Division. Their run was from Tacoma and Seattle on Puget Sound, to Othello, which lay 214 miles east in farming country. Their primary work was pulling regular passenger trains. They did, however, handle silk trains in the early days, and occasional excursions. In the 1930's and 1940's, they also pulled ski trains to the Milwaukee Ski Bowl in Snoqualmie Pass.

During their stay on the Coast Division, crews considered them to be truly great locomotives. Looking back, retired engineer, Ralph Edwards recalls:

> "They were quick. If you reached over and grabbed a handful of notches those babies would get right out and move. The acceleration was much quicker than a diesel and much quicker than E22 and E23. They moved out in a hurry and they were absolutely silent. With twenty-four drivers they had good adhesion too. The visibility out of the front wasn't very good, but it was no worse than on a steam engine."

Retired engineer, King Clover, also remembers their quick accelerations:

> "My first experience with the Bipolar's acceleration was in 1941, when I was a hostler," says Clover. "Me and a partner of mine were listening to an old timer's stories about how fast the Bipolars would go. Since we didn't believe everything he told us, the old guy made us a bet. He told us to stand by the front pilot and he bet us he could accelerate so fast that we couldn't jump on the back end when it went by. We took him up on the bet and we tried to jump on, but we just couldn't do it."
>
> "The Bipolars were designed to operate in the Cascade and Saddle Mountains," says Clover. "They were a good mountain engine and they ran fine at speeds as high as 60 to 65 mph. On the Coast Division there were very few places where you could exceed that. The Bipolars were an ideal engine if you operated them the way they were designed to be run."

One facet of Bipolar design which was surprising was the amount of flexibility in their unique four-truck chassis. The Bipolars, like all Milwaukee Road power were designed to accept the ten degree curves which were common on the mainline. A rigid wheel base of only 13 feet 9 inches made that easy, but such curves did not approach the Bipolar's limits. The limits were far beyond the expectations of even the crews who ran them. King Clover discovered this

E4 at Tacoma Yard being prepared for a run. The train heat boiler is hot and the wheels are getting a hot water wash. 6-3-50./Photo by Albert Farrow.

when he rounded a curve in the Cascades and discovered trouble in the path of his train.

"A tree had come down at Hyak and knocked the track twelve inches out of line in the shape of a "u." I knew we couldn't stop before we got there and I thought we were going over in the ditch. Instead we went over that track just fine and the Bipolar stayed on the rails. By the time we got stopped, the out of line track was under the baggage car."

According to Ralph Edwards, "The Bipolars were a fine piece of equipment, but they had their peculiarities. When they got too much lateral play in their driving boxes, they just beat you to death when you got on rough track. Prior to World War II the Milwaukee did a pretty good job of maintaining them. During the war they were pressed for power, just like any other railroad and they only did what they had to as far as running repairs. The Bipolars wre designed to handle nine coaches, but during the war they would hang fourteen cars on them over the Beverly Hill. That burned them up due to overloading. They were running two passenger trains each way every night and sometimes they would be double headed. They would double head from Tacoma to Beverly. There, one Bipolar would cut off and wait for the train coming the other way to help it over the hill. There were troop trains too. Those motors were on the go almost 24 hours a day. During the war they used anything they could get to move trains except a switch engine. Sometimes they would double head steam and Bipolars. Most often it was the mallets if we could spare them or the L Engines (2-8-2's). The Coast Division didn't have any passenger steam engines assigned to it. The only time I recall seeing an S Engine (4-8-4) over here is when they were being sent to the backshop in Tacoma."

"B" end view of E1 from the fireman's side. Tacoma Yard 4-6-51./Photo by Albert Farrow.

Among the Coast Division crews there is seldom a negative word about Bipolars unless the discussion turns to the train heat boilers. Then the criticism begins.

> "The Bipolars were a fireman's nightmare," says Ralph Edwards. "They were designed for nine coaches and when the railroad started adding more than that, a fireman really had problems trying to keep steam up. The boiler just wasn't big enough."
>
> "They were the toughest boilers to fire successfully that I ever saw," says King Clover. "I spent some smokey nights fighting that sucker. Just about the time you would start patting yourself on the back, saying, boy I sure know how to fire these boilers, that's about the time something would go wrong. You would spend the rest of the trip in the boiler room."

People who watched the Bipolars from trackside knew nothing of troublesome boilers. It was bells and whistles which caught their attention; the sounds which warned of a noiseless train near the grade crossing.

The bells were rung more often than bells on other types of motive power and sometimes the bells rang continuously. The non-stop ringing was caused by an air ringer valve which tended to stick in the "on" position. Train crews were not anxious to loosen it during a trip because there were too many high voltage wires near its location on the hood.

Bipolar whistles were remembered because of their high pitched, almost European sound. The sound contrasted sharply with the low pitch of American steam locomotive whistles. The Bipolars retained these whistles until the late 1940's when airhorns were installed to replace them. The sound of the whistle tended not to carry for long distances and the addition of

The walkway inside the #2 end hood of a Bipolar was lined with 3,000 volt circuits and control equipment./Photo by General Electric.

E5, E4 and train race southbound from Seattle. Date unknown./Photo from Warren Wing Collection.

*In 1948, E1 became the first of the Bipolars to be repainted in Hiawatha colors. She had orange ends with silver stripes, an orange lower body, an orange stripe below the road name, maroon side panels, and a dark gray upper body. The trucks were black. This paint scheme was not duplicated on any of the other Bipolars. Location Tacoma yard. Date unknown./*Photo courtesy Milwaukee, Wisconsin, Public Library.

horns was a safety measure. Sealed beam headlights were also added at this time in order to make headlight maintenance simpler. Some Bipolars received three bulbs in each headlight shell and others received four.

For the first 30 years, the Bipolars' paint scheme remained a basic black. Occasionally they were given white trim, but no colors were applied until after the arrival of the steamlined Olympian Hiawatha.

Like other Hiawathas of the 1940's, the Olympian's cars were orange with a broad maroon band at window level. Diesels and steam engines used for pulling Hiawatha trains in the midwest were painted maroon and orange to compliment their trains, but no standard pattern existed. In 1948, E1 was given orange ends, silver stripes, maroon side bands and a dark gray top. It was a paint job inspired by the pattern on two of the Fairbanks-Morse diesel sets assigned to the Olympian Hiawatha. In 1949, E3 was painted in a different Hiawatha pattern. This one had a maroon band around the locomotive and silver wings on the nose. It resembled the Fairbanks-Morse diesel sets which had chrome plated wings on their fronts. E3 was the Bipolar displayed at the 1949 Chicago Railroad Fair. After its return, the silver wings pattern became standard for all but E1. It remained so until 1952 when a pattern identical to the one used on Little Joes, and 1950 era diesels was adopted.

By 1952, the Bipolars had over thirty years in service. They were technologically obsolete and due to poor wartime maintenance, their condition was not the best. Their next stop could have been the scrapper but the railroad failed to purchase enough new electrics to replace them. As a result, Electrification Department Head, Laurence Wylie set to work designing improvements for the Bipolars. They were a group of locomotives which he praised for their ruggedness and reliability.

Wylie was a top notch electrical engineer and some labeled him a genius. Ralph Edwards' recollections sum up those of many other railroaders when he says:

> "There was nothing about a Bipolar that Laurence Wylie didn't know, and if he didn't have an answer he would get out his slide rule and come up with an answer damn quick."

Wylie's answer was to provide: additional traction motor shunting for increased speed, better traction motor ventilation for more horsepower, roller bearings for high speed reliability, MU controls so that one crew could operate two locomotives, improved guiding truck centering for better tracking, high capacity flash boilers for reliable train heating, and streamlining to make them look new.

E5 was the first to get these improvements. It was rebuilt at the Tacoma shops under Wylie's watchful eye. Upon completion, E5's performance was everything Wylie said it would be, but the price tag was above his estimates. Modifications would cost between $35,000 and $40,000 per locomotive. The Chicago office granted permission to proceed with the project, but sought a lower cost.

Two rebuilt Bipolars prepare to leave the new Tacoma Station for Seattle. Date unknown./Photo by Laurence Wylie.

Bipolar E1 was a recent arrival on the Rocky Mountain Division when this photo was taken on May 2, 1957. It is preparing to take the Olympian Hiawatha west out of Deer Lodge./Photo by R.V. Nixon.

E5 at Deer Lodge, Montana May 2, 1957./Photo by R.V. Nixon.

Shops were the domain of the Mechanical Department, and the Mechanical Department successfully argued that any project this large could be more efficiently carried out by the main shops in Milwaukee, Wisconsin. The large, well equipped Milwaukee shops were famed for their construction of steam locomotives, passenger cars and freight cars. Using E5 as a model, they rebuilt the other four Bipolars during 1953, and turned them out with an appearance even better than E5. Unfortunately, appearances are not everything. The Milwaukee shop crew was inexperienced in working on Bipolars and high voltage electrical systems. The results of their work were depressing to Wylie.

"They did a poor job. Thereafter, these passenger motors were plagued by electrical fires, armatures coming apart, and other problems. We didn't use them much after that," he said.

Barry Kirk, who headed the Electrification Department from 1963 to 1971, was one of Wylie's assistants at the time. He recalls walking through one of the Bipolars when it first arrived as a dead-headed unit from Milwaukee.

"I saw a clump of wires coming out of a cabinet and there was a tag on them. It said, 'These wires are hot but we don't know where they go.' "

The Tacoma shops attempted to remedy the problems but their efforts to make the Bipolars reliable were not a total success. The problems were varied and unpredictable. Some resulted from poor electrical work in Milwaukee, others were due to hard use and the age of the locomotives.

Steam drifts from the train heat lines of E4 as it waits in Deer Lodge with the eastbound Olympian Hiawatha. It is April 29, 1958 and still cold enough that some snow remains./Photo by Doug Cummings

E1 in Union Pacific colors at Deer Lodge, April 29, 1958. She was trouble-prone on the fast rolling Rocky Mountain Division, but she still looked grand./Photo by Doug Cummings.

The optimistic Electrification Department continued to hope that the next repair job would solve all remaining problems. Worried dispatchers, however, sometimes assigned two Bipolars to a train that only required one. In this way they assured on-time arrival even in the event of locomotive failure.

From 1954 through 1957, the use of Bipolars decreased in favor of more reliable diesels. The Bipolars were often held in reserve even after being declared fully fit for service.

> "After they stopped using the Bipolars in passenger service out here, they talked about using them for freight, but it wasn't something they seriously considered," says King Clover. "I did run one on a work train for a rail laying gang, but it was terrible for that. You were always running in resistance or with half of the motors cut out to make them run slow enough for that service."

In order to help phase Little Joes E20 and E21 out of passenger service, the Bipolars were sent to the Rocky Mountain Division in mid-1957. The Deer Lodge shops repainted them in mid-1958 to conform with the Milwaukee's new, Union Pacific inspired, yellow and gray passenger paint scheme. The Railroad adopted these colors after acquiring the contract to haul Union Pacific streamliners between Omaha and Chicago. It was proud to provide this service and thus adopted the U.P. colors system-wide for its own passenger trains.

When the Bipolars entered Avery to Harlowton service, their performance was not up to the expectations of Rocky Mountain Division personnel, and they were not warmly received. The Joes and the Westinghouse Motors were smooth riding locomotives which were fully capable of hauling passenger trains in excess of 80 mph.

Although General Electric claimed in 1919 that the Bipolars were capable of 90 mph, they had never run that fast on the Coast Division. Additionally, their ride was never characterized as smooth in high speed operation.

> "At 60 mph and under, they rode good, but if you got them up over that, they would wiggle like a snake, ride rough and flash over their traction motors," says King Clover. "Between Avery and Harlowton they were kind of a beast due to excessive speed. When the Bipolars were first sent to the Rocky Mountain Division, I was a Road Foreman of Engines. The Milwaukee sent me over there to show the engine crews how to run them. The first run I made on the Rocky Mountain Division was between Deer Lodge and Missoula on #15. We had a double Bipolar and I was riding in the second unit. When we went through the first tunnel, the floor boards flew up off the cab floor. I said to myself, 'My God, what's going on here.' I looked down at the speedometer and saw we were going 85 mph. After the trip was over, I talked to the Master Mechanic in Deer Lodge. I told him that they would have to lower the speed limit to 60 mph if they wanted the Bipolars to last. The Master Mechanic just shook his head and said, 'We just have to go to make the schedules.' Their schedules over there were so tight that they didn't have any slack."

The Bipolars quickly developed a reputation on the Rocky Mountain Division for rough rides and unreliability. Retired Rocky Mountain Division fireman Bill Merrill describes their ride as being like that of a lumber wagon. Other crews dubbed them Polar Bears, Centipedes, and Caterpillars. Because of the problems which developed, the Milwaukee reduced the Bipolar speed limit to 60 mph. Still they did not last long. Due to persistent problems they were pulled out of service one by one starting in 1958. By 1960, all were in storage.

Three Bipolars are towed their last mile to a scrap yard in Seattle. Soon they will resemble the formless junk in the gondola they follow./Photo by Robert Oestriech.

"The thing you have to think about," says Clover, "is that the Bipolars were on the Coast Division forever and we liked them. On the Rocky Mountain Division they hated Bipolars like the dickens. The railroad told them they would have to run Bipolars and there were engineers who did everything they could to get rid of them."

After being withdrawn from service the Bipolars sat in Deer Lodge awaiting disposition. In 1962, E1, E3, E4, and E5 were towed to Seattle and scrapped. E2 was donated to the St. Louis Museum of Transport. E2 entered service in 1918 as the 10251. At the time it was donated, it was painted in the yellow and gray Union Pacific streamliner colors which the Milwaukee was using in 1962. It is displayed out of doors and when weathering brought about the need for a new paint job, the museum applied the more traditional maroon and orange.

Although the Bipolars were 43 years old and technologically obsolete when scrapped, two passenger diesels were needed to provide pulling power equal to one Bipolar. What is more the diesels were unable to match the fabled acceleration of their predecessors.

Inside the cab of Bipolar E2 photographed at the National Museum of Transport June 28, 1984./Photo by Donald R. Kaplan.

Bipolar boiler room minus the boiler. National Museum of Transport June 28, 1984./Photo by Donald R. Kaplan.

THE WESTINGHOUSE MOTORS
CLASS EP-3

These ten locomotives were the longest, tallest, heaviest and most troublesome group of electric motors the Milwaukee ever owned. They were loved by many Rocky Mountain Division operating crews for their smooth ride, speed and power. They were also hated by many maintenance men and company officials for their persistent and expensive to fix problems. Like the GE Freight Motors, they had many names and except for the designation EP-3, none was officially considered more proper than the others. Depending on who you talked to, they were the Westinghouse Motors, the Quills or before they were renumbered, the Ten-three-hundreds.

Westinghouse was not the Milwaukee's first choice as a locomotive builder. However, in 1917 the Milwaukee was only one of many railroads which did not get its first choice. The United States entered World War I on April 2, 1917, by declaring war on Germany. In order to assist the marshalling of national production, the United States Railroad Administration took control of the railroads.

Among other things the USRA standardized steam locomotive designs and designated who would build them. Electric locomotive designs could not be standardized when so few railroads were electrified, but the USRA utilized its authority in directing the Milwaukee's locomotive purchase. The Milwaukee proposed purchasing a new group of fifteen, state-of-the-art, 265 ton passenger motors from General Electric. The USRA responded that in the interest of utilizing the national production capabilities more effectively, ten of the fifteen must be manufactured by Westinghouse.

Westinghouse designed the EP-3's to meet the same performance specifications as the Bipolars. However, they embodied an entirely different approach to locomotive design, an approach influenced by steam locomotive designers.

The World War I era standard in passenger steam engines was a high wheeled 4-6-2. The Westinghouse Motors were two 4-6-2's back-to-back, complete with high wheels. The long tall chassis and box shaped cabs to house the electrical gear were built by Baldwin Locomotive Works, America's leading builder of steam engines, and a subcontractor for Westinghouse.

At the time Westinghouse received the contract to build passenger motors for the Milwaukee, it was already involved in building some for the New Haven. In many ways the group built for the Milwaukee was an enlarged version of the New Haven motors. The changes were designed to reflect the size and power requirements of mountain railroading.

The New Haven EP-2 Class Passenger Motors delivered in 1919 had a 1-C-1 + 1-C-1 wheel arrangement, rode on 63 inch drivers and had a 3.2 to 1 gear ratio. The larger Milwaukee Motors were a 2-C-1 + 1-C-2, rode on 60 inch drivers and had a 3.7 to 1 gear ratio. Both used twin armature traction motors driving the wheels through spring cushioned quills.

The first two locomotives in the Milwaukee's Ten-three-hundred series were delivered in December of 1919. The 10300 and 10301 immediately overshadowed their New Haven predecessors with their immense size and weight. They were 88'7'' long, 17' high and weighed 283 tons. Rated at 3,400 horsepower continuously, they had more than 1½ times the capacity of their East Coast sisters. In addition, they had more power and weight than the Bipolars, which GE delivered a year earlier. The "Quills" as Westinghouse dubbed them, were immediately the world's largest electric locomotives. Based on a year of testing at the Westinghouse plant in East Pittsburgh, Pennsylvania, Westinghouse expected the exploits of

Builders photo of Westinghouse Motor 10300. Note the fragile looking frames and the quill springs on the wheels. This motor became E10 in 1939./Photo courtesy Milwaukee Road.

these boxy but businesslike passenger motors to take the railroad world by storm. Mathematical models developed by the company indicated that a single Quill could pull almost as much as two USRA 4-8-2's.

These huge passenger motors were designed to have a top speed of 65 mph, and a starting tractive effort of approximately 102,000 pounds. The greatest over-the-road pulling power was produced at 24 to 26 mph, the speed at which a passenger train would ascend a mountain grade. These locomotives could draw up to 3,300 amps for one minute while starting, but in normal operation they drew only 945 amps.

A Westinghouse Motor was bi-directional and had operators' cabs at each end of its body. Connecting the cabs were aisleways going down each side, and between the aisles sat electrical gear. In the center of the locomotive between the aisles was an oil-fired vertical boiler and two water tanks. Below the boiler and beneath the floor was a fuel oil tank.

The locomotive rode on an articulated chassis: two sections linked by a drawbar. The chassis was made of steel bars and carried six 566 horsepower traction motors, one mounted directly above each of six driving axles. Each traction motor had two armatures and the armatures were geared to a quill.

The quill was a steel tube 15 inches in diameter which was mounted around the axle. On each end of the quill were "spiders" with seven arms. These arms were linked to the seven driver spokes through coil springs. The springs were held in place by cups on the driver spokes and the spider arms. If the traction motors were to turn the quills suddenly, the springs would compress and absorb the jolt before passing the power on to the drivers.

The reason for this complicated cushioning system was concern about the enormous power of the traction motors. Westinghouse design engineers felt certain that without cushioning, a 566 horsepower traction motor would cause mechanical damage to the locomotive during sudden starts. Unlike a steam engine, an electric motor can exert full horsepower from a standing start. Design engineers were unfamiliar with such power in 1919.

Auxiliary equipment on the Westinghouse Motors included an 85 volt motor-generator set and two 85 volt axle generators. The M-G set used a 50 horsepower, 3,000 volt motor to turn a

Front end c E11 in 1940. hit an autc mobile west c Missoula. Photo by H.I Morgan, Go don Rogers Co lection.

10300 at Three Forks on August 9, 1938. Numerous changes have been made in the 20 years since she was built./Photo by Otto Perry, courtesy Denver Public Library.

Photos of Westinghouse Motors on the Coast Division are rare. In this one, the 10309 and train await a highball at Seattle's Union Station./Photo by Asahel Curtis, Jim Fredrickson Collection.

35 KW, 85 volt generator. This unit provided current for train lighting, locomotive control circuits, and battery charging. The axle generators were rated at 40 KW each. They were connected to one axle of each pilot truck and operated at any speed above 12-15 mph. They provided current to operate the air compressor, and the traction motor blowers as well as current for exciting the traction motors during regenerative braking.

All of the Westinghouse Motors were on the property by May of 1920. They entered service painted black with white lettering. Road numbers were 10300 through 10309. They kept these numbers until March of 1939 when they were renumbered E10 through E19. The renumbering was not quite in sequential order because the 10303 had been wrecked and was in storage awaiting disposition. The 10303 would have become E13, but since its future was in doubt, it was assigned the number E19. All of the others were moved ahead in the series. In addition to new numbers, these passenger motors were given white stripes which they carried until approximately 1950. At that time they were painted in the orange and maroon which they carried until scrapping.

Between 1920 and 1950, this group of passenger motors was rebuilt at least five times. Some work was done in Deer Lodge and some in Tacoma. Also, many locomotives were rebuilt separately. As a result each locomotive took on a slightly different appearance. Vents and ducts varied in size and shape. Some locomotives had fairings and air deflectors, others did not. Horn locations varied, and some front doors swung in, while others swung out.

One change did, however, become uniform. During the 1930's the Westinghouse Motors became single ended locomotives. This was accomplished by simply removing the controls from the #1 end. Little is remembered about the reasons for this, but years later the Milwaukee did the same with the Little Joes. Dual controls were expensive to maintain and little need for them was seen. All direction changes were being made in terminals such as

#10301 after being rebuilt with a 2-C-2+2-C-2 wheel arrangement and a two-unit superstructure. Date probably 1927 or 1928./Photo courtesy Milwaukee Road.

10301 at Butte, Montana, September 26, 1931. She is once again a single-unit motor of standard appearance./Photo by Otto Perry, courtesy Denver Public Library.

Westinghouse Motor E10 with The Olympian at Deer Lodge 4-6-40. The front pantograph has a wind fairing around its base./Photo by R.V. Nixon.

Harlowton, Deer Lodge, and Avery. These terminals had turntables and the locomotives could be turned during layovers when they were being serviced.

Over the years various pieces of equipment came to be installed in the vacant cabs of different locomotives. In the E14, it was a large tank of fuel oil which crew men had to squeeze past to get out the back door.

Some of the huge and speedy Westinghouse Motors lasted for 38 years. During that time, their history was etched in the memories of those who worked with them. They were controversial with notable strong points and weak points.

When these passenger motors entered service, engine crews were impressed. It quickly became apparent that in many ways these passenger motors were an engineer's dream. They rode smoothly at speed and could easily pull a passenger train at 70 to 80 mph. They could pull more tonnage than the spec book said they could, and they didn't have the wheel slip problems of the older GE passenger motors, the EP-1's. Opinions were not unanimously positive, however. There were also a number of problems.

According to retired fireman Bill Merrill,

> "The biggest problem with the Westinghouse when they first came was that they would break their frame. The second big problem was the quill springs. The springs were about two feet long and about eight inches in diameter. Those springs weren't fastened on too good and when you got going 70 to 80 miles an hour you might hear them go whistling through the right-of-way fence." "Electrically the Westinghouse was OK, but they was tricky to work on. Very few electricians could maintain them."

The passenger motors Westinghouse built for the New Haven never had frame problems, but a comparison of the locomotives built for the Milwaukee and the New Haven shows a startling difference. The 175 ton New Haven motors had heavy cast steel frames while the 283 ton Milwaukee version had light frames made of steel bars.

Westinghouse built the New Haven class EP-2 passenger motors in 1919, just before the Milwaukee's EP-3's were built. The New Haven motors were 69 feet long, weighed 175 tons and had heavy cast steel frames. They had no problems with frame breakage./Photo by Westinghouse.

When broken frames began appearing on the Milwaukee, Baldwin Locomotive Works almost immediately built heavier frames for the big Baldwin-Westinghouse products. This, plus heavier Delta idler trucks in the center increased the locomotive weight to 310 tons. Though the new frames were built of very heavy steel bars, this did not entirely solve the problems.

> "Side frames had to be replaced a total of three times," said Laurence Wylie. "The frames tended to crack or break as a result of derailments, and these locomotives were involved in a lot of derailments."

Wylie felt tall drivers were a contributing factor in the broken frames. Because of their size they could exert a lot of twisting force if derailed. In critiquing his least favorite electric, Wylie pointed out that driving wheels five feet in diameter are unnecessary on an electric locomotive. Unlike steam engines, gear driven locomotives can achieve their speed through their gear ratios.

Another of the frame problems possessed by the Westinghouse Motors was stiffness. The rigid wheel base of these 310 ton locomotives was a very acceptable 16'9''. That statistic, however, does not tell the whole story. The non-rigid wheel base was not very flexible. It was so inflexible, in fact, that the operators manual states,

> "THESE LOCOMOTIVES ARE NOT TO BE TURNED ON A WYE unless you have orders from the dispatcher to do so."

Those are ominous words when you consider that the Milwaukee's mainline had ten degree curves.

Adam Gratz, a retired engineer was a fireman back in the days when Westinghouse Motors were running. He occasionally had to help wye them and recalls it as a very touchy process:

> "You couldn't run them through a wye at speeds any faster than a walk, and you had to watch all of the wheels closely. The Westinghouse frames were so stiff that the wheels were liable to lift up over the railhead and derail."

Another problem caused by the stiff frames was wheel flange wear. Since the wheels had limited lateral movement, there was grinding action between the wheels and rails as the locomotive passed through curves. In an experimental effort to reduce the flange wear, Baldwin rebuilt the 10301 during the mid 1920's. They produced a permanently coupled two unit locomotive with a 2-C-2+2-C-2 wheel arrangement. This did not, however, solve the problems of stiff frames and flange wear.

After at least a year of testing, some of it on the Coast Division, the 10301 was returned to its previous appearance. The Westinghouse fleet continued to spend more than its share of time in the shops, having new steel tires mounted on the cast steel wheel centers.

Many shop workers considered the Westinghouse Motors to be a nightmare to work on. Retired Electrification Department Head, George Frazier, explains some of the reasons:

> "They were full of tight cramped places that you had to get into in order to work on them. For instance changing the traction motor brushes was hard to do. The electricians would have to crawl between the frame and the body to get at them. It was always so greasy and dirty down there, that the electricians would have to wear rain gear just to try and stay clean."

The Westinghouse Motors spent almost all of their career on the Rocky Mountain Division. Occasionally, however, they ran on the Coast Division. According to Laurence Wylie, maintenance problems kept them from staying long on the Coast. Coast Division roundhouse crews did not know how to maintain them.

BIPOLAR/WESTINGHOUSE COMPARATIVE TONNAGE RATINGS
COAST DIVISION, 1936

EASTBOUND	RULING GRADE	EP-2 BIPOLAR	EP-3 WESTINGHOUSE
Tacoma-Black River	0.0%	3450	4300
Black River-Cedar Falls	0.8%	1715	2200
Cedar Falls-Hyak	1.74%	860	1050
Hyak-Cle Elum	down	3000	3500
Cle Elum-Kittitas	down	3000	3500
Kittitas-Boylston	1.6%	925	1150
Boylston-Beverly	down	none listed	none listed
Beverly-Othello	0.4%	2300	2500

****

WESTBOUND			
Othello-Beverly	down	3500	3500
Beverly-Boylston	2.2%	580	600
Boylston-Kittitas	down	1400	1500
Kittitas-Cle Elum	0.4%	3500	3600
Cle Elum-Hyak	0.7%	2400	2500
Hyak-Cedar Falls	down	1250	1350
Cedar Falls-Black River	down	3500	3500
Black River-Tacoma	0.0%	3500	3500

NOTE: Bipolars and Westinghouse Motors had only one air pump and thus were limited to 50 cars in freight service. Longer trains required two air pumps for safe train braking.

10304 at Butte, Montana, September 26, 1931./Photo by Otto Perry, courtesy Denver Public Library.

10305, the motor that President Harding ran, sits at Deer Lodge on August 9, 1938. The #1 end has been blanked out and squared off. She also has non-standard vents, etc./Photo by Otto Perry, courtesy Denver Public Library.

10307 at
Forks, M
1934./Pho
Otto P
courtesy D
Public Lib

E10 at Butte, Montana with the eastbound Columbian on a cold winter night in 1957./Photo by Richard Steinheimer.

Retired electrical engineering consultant Walter Gordon rode one of the "Quills" on the Coast in 1927 and spoke of his memories some fifty years later:

> "I was working for the Milwaukee then," said Walter. "I was checking the trolley voltage at various points along the line from inside the 10307 or 10308. She didn't have some of the modifications that the Rocky Mountain Division Quills had, her internal equipment wasn't where the operator's manual said it should be, she had a rough ride and always needed maintenance on those quills. She was a clunker."
>
> "The reason she was out here," said Walter "was because of the extra passenger trains and excursions the Milwaukee was running. There were specials to Spokane, excursions to Mt. Rainier which were electric from Seattle to Tacoma, and silk trains. Sometimes the Milwaukee would need six passenger locomotives at once on the Coast, but there were only five Bipolars. Since nine of the Quills could handle the Rocky Mountain Division, they sent one out here."

It is interesting to note that Westinghouse Motors are mentioned in the 1926 employees timetable for the Coast Division. Special instructions limit passenger trains pulled by Westinghouse Motors to no more than 20 mph while rounding McClellands Butte curve in Snoqualmie Pass. This timetable also lists the weight of an EP-3 as 276 tons. The Rocky Mountain Division timetable issued the same year lists the EP-3 as a 307 ton locomotive.

An impressive feature on the Westinghouse Motors was their regenerative braking. They could enter the braking mode at almost any speed and this allowed excellent train control in the mountains. The motors which General Electric built could only enter regenerative braking at certain speeds. As train crews made use of this flexibly useful braking, they discovered an unexpected fact. Through improper use of the braking system, ballast scorching throttle jerkers could roll their trains at speeds that were unheard of in the Rocky Mountain West.

According to retired fireman Bill Merrill,

> "There was one thing you could do on a Westinghouse and make it beat anything on wheels. You see, the Westinghouse had two shunts per speed where a GE only had one. A shunt would weaken your traction motor field and give you more speed but less pull. You could run a Westinghouse wide open in the second shunt and then put

Westinghouse Motors E12 and E14 at Soudan with the Olympian on 9-12-39. Two crews were needed because these motors were not equipped for MU operation./Photo by R.V. Nixon.

The first section of #15, the westbound Olympian, pulls out of the siding at Bonner Junction in August 1943./Photo by R.V. Nixon.

> it into regeneration to weaken the field even more. It was almost like another shunt. Now the motor was set to regenerate and slow down as you closed the throttle. Instead though, you would leave the throttle wide open and keep motoring. Then you would get going faster instead of slower. You could get your train up to 90 miles per hour on level track like that. It was strictly against the rules though. It was hard on the traction motors and might create a flash-over.''

Experience repairing flash-over damage led the Milwaukee to include a very specific rule in its revised EP-3 operators manual.

> ''The use of regeneration for the purpose of shunting the traction motor fields through the axle generators in order to increase the speed of the locomotive is strictly prohibited.''

For many years, the man enforcing this rule was Rocky Mountain Division Master Mechanic Bill Brautigam.

> ''Brautigam was a pretty smart guy,'' says Merrill. ''You could flash over a traction motor and he could tell what you had been doing wrong even though he wasn't there.''

Another retired fireman with memories of the Westinghouse Motors is Ralph Danley. He spent most of his career working with steam engines on the Great Falls branch. On occasion, however, he was called from the extra board to fire electrics on the mainline. Most of his memories involve the train heat boilers he had to tend.

> ''The boilers were a little small and had trouble producing enough steam to heat 12 or 13 cars. If the temperature dropped to 30 below a fireman really had to work to keep his train warm.''

Westinghouse E13 in the Bitterroots above Avery. Date unknown./Photo courtesy Milwaukee, Wisconsin, Public Library.

When the fireman was raising steam in a cold boiler, a compressed air jet provided the draft to take smoke up the stack. If this was not working properly, the boiler could belch oil smoke inside the locomotive.

The boiler fuel was another problem. The low grade tar-like bunker oil used as fuel would not flow well until heated to 120 degrees. Additionally, it did not burn well unless pre-heated to a temperature between 170 and 180 degree. If cold, it would just lie there in the bottom of the firebox, producing little heat.

Firemen had to watch the oil tank temperature closely. According to Danley,

> ''The oil would congeal into a liver-like mass if it was heated above 185 degrees.'' Congealed oil wouldn't flow through the boiler's atomizer. The fireman would have to just scoop that liver-like junk out of the fuel tank and fill it with new oil. Scooping the oil out was a lot of work. One time Red Pryor cooked his oil so bad that it couldn't be scooped out. They had to cut the end out of the oil tank to get the oil out. It was a cold day and he forgot to turn off the steam heater in the oil tank after his oil got warm.''

Bunker oil tends to make a lot of soot when it burns. Among the duties of the fireman was blowing the soot out of the flues once an hour. Soot in the flues could reduce the efficiency of the boiler and sometimes choke the fire by trapping smoke in the firebox. These boiler problems were never really solved until the 1950's. At that time "flash boilers," similar to those used in diesels, were installed.

From 1919 until 1950, the Westinghouse Motors were the primary passenger motive power on the Rocky Mountain Division. Occasionally when none were available, steam engines or steam double-headed with GE Freight Motors were used, but this never occurred very often.

The Westinghouse Motors were designed to pull passenger trains singly and did so for most of their career. They were double-headed only when trains were unusually long. Such trains were sometimes generated by World War II troop movements. Since these motors were never equipped with multiple-unit controls, double-heading brought on the additional expense of extra train crews.

The Westinghouse Motors had tonnage ratings for freight service as well as passenger service, but no serious consideration was given to using them for freight. They were looked upon as acceptable only for short, fast, freights of 50 cars or less. With the exception of silk trains in the early years, the Milwaukee preferred to run long, heavy drag freights which moved considerably slower than passenger train speeds. Weak frames, lack of MU controls, lack of airbrake capacity, and incompatibility with the GE Freight Motors precluded the use of Westinghouse Motors on long, heavy freights. Such freights would have required the tractive effort of two or three Westinghouse Motors each.

In the years following World War II declining passenger traffic and changes in motive

No thunderous sounds, no pillar of smoke, just a loud hum and some gear tooth growl as E17 glides up the 2% grade at Vendome with the ten car Olympian on 3-14-46./Photo by W.R. McGee.

End door of 10305 after being modified to open outward in 1938./Photo by H.R. Morgan, Gordon Rogers Collection.

power utilization meant a reduced role for Westinghouse Motors. In 1947, the Spokane to Butte local was discontinued. The streamlined Olympian Hiawatha entered service that year, but even in electrified territory it was pulled by Fairbanks-Morse diesels. Due to a coal strike in 1949, which idled many of the Milwaukee's steam engines, the oil fueled Fairbanks-Morse diesels were needed on other parts of the system. On the Rocky Mountain Division, the train was temporarily assigned to Westinghouse Motors. In 1950, Little Joes E20 and E21 entered service and were given the assignment.

In 1952, Laurence Wylie made plans to rebuild the Bipolars and GE motors for further service. The Westinghouse Motors were, however, purposely excluded from his plans. Because he disliked them, they were slated to be phased out. At that point in time, only seven of the ten Westinghouse Motors were still in service. E19 (10303) had been involved in a wreck in

1933 in Avery. It was stored in Deer Lodge for nine years following the wreck. In 1942 it was consigned to the scrappers as part of the war drive to gather steel. When moved from its storage spot, it fell into the Deer Lodge turntable pit and was ignominiously scrapped on the spot. E13 (10304) was demolished at Soudan in 1947 when it hit a mud slide at speed. E17 (10308) was heavily damaged by fire in 1950 and scrapped.

Through the use of the two Little Joes, the two streamlined, regeared GE Passenger Motors and later the Bipolars in Rocky Mountain Division passenger service, it was possible to retire the remaining Westinghouse Motors. They were retired as they came to need heavy repairs. E14 (10305) in 1952; E12 (10302) in 1954; E10 (10300), E15 (10306), and E16 (10307) in 1955. The last two remained in service until 1957 when it was decided that there were too few remaining to justify stocking parts. These last two were E11 (10301) and E18 (10309).

Many engineers and firemen were sorry to see the Westinghouse Motors retired. The smooth ride and power made them favorites. The Bipolars, which initially replaced them had a rough ride and experienced numerous problems when operated above 65 mph. It was a common practice on parts of the Rocky Mountain Division to ignore the track speed limits in order to make up time. Since the Bipolars could not easily do this, crews compared them unfavorably with the Westinghouse Motors. Many would agree with retired engineer Swede Hansen when he said:

> "The Westinghouse was almost as good as a Joe, if the Westinghouse wasn't overloaded. When they scrapped the Westinghouse and gave us the Bipolars, they scrapped the wrong motors."

In contrast, when Laurence Wylie looked back on the history of the Milwaukee's huge quill driven speedsters, the memories were different:

> "The Westinghouse Motor was not a good locomotive," he said. "They ran well when they were running, but due to one problem or another they weren't always running."

Westinghouse Motor E10 is hitting her stride with train #18, the Columbian, near Alberton in 1953./Photo by Jim Fredrickson.

THE LITTLE JOES
Classes EP-4 and EF-4

The Little Joes were powerful, rugged and fast. They were the Milwaukee's most modern electric locomotives, and their excellent performance made them legends in their own time.

They were built for Joe Stalin's Russia and when they went instead to the Milwaukee, an unknown employee named them Little Joes. That name was adopted almost universally but was sometimes shortened to Joes. Except in official documents, it superseded such names as GE 750 type and the class titles EP-4 and EF-4.

Immediately following World War II, the Russians made plans to extend their 3,000 volt electrification across Siberia. Locomotives were needed for this project and the Russians inquired about having General Electric build them. The U.S. was a wartime ally, and G.E. was a respected name on the Soviet Railway. In 1929, G.E. had delivered a group of 3,000 volt C-C locomotives which the Russians considered good enough to use as models for a fleet of home-built duplicates.

The Soviet Railway, SZD, was built to the uniquely Russian gauge of five feet between the rails. The Trans-Siberian route was one of gentle curves, none of which exceeded five degrees. Its maximum grades were 1.5% and train speeds were expected to regularly reach 60 mph. The use of high capacity substations and a voltage which was 10% above the old standard would guarantee sufficient current for the utilization of large locomotives in heavy service. In the months following V-E Day in 1945, design engineers from General Electric reviewed the route specifications. Before the year was out, their proposal was drawn up and submitted. They proposed building a custom designed locomotive in order to provide optimum performance. The new locomotive would be a 4,750 hp streamlined, dual cab, 2-D + D-2 which was designated the GE 750 type. The Soviet Ministry of Railways was apparently pleased with the proposal, because in March of 1946 an order was placed for 20 prototypes.

The locomotives were constructed at the G.E. plant in Erie, Pennsylvania, and in May of 1948 the first of the new locomotives was completed. Unfortunately for the Soviet Railway, East-West relations had rapidly soured after the conclusion of the war. Russia began cementing its hold on territorial gains in Eastern Europe, and in 1947 the U.S. retaliated by providing military assistance to nations threatened by Communist takeover. A cold war of nerves began as the Russians blockaded Berlin in 1948 and the U.S. airlifted supplies over the roadblocks. President Truman, who was strongly anti-Communist, then banned the sale of all strategic materials to Russia. Both the U.S. and Russia considered railroads to be of strategic importance and thus General Electric was suddenly stuck with 20 locomotives which were unpaid for and undeliverable. The locomotives which were now coming off the assembly line at the rate of two or three per month were tested and then stored behind the factory.

Because of their specialized design, the new locomotives had a very low value to any railroad other than the SZD. Although the gauge of the wheels could be changed, the locomotives were too large and heavy for use on most European railroads, and they required more current than most existing 3,000 volt substations could provide. The U.S. market appeared as bleak as that in Europe. There were only two 3,000 volt systems using locomotives in the country and neither wanted new electrics. The well-equipped, 17-mile long Cleveland Union Terminal needed nothing, while the Milwaukee's electrification was facing abandonment. Having nothing to lose, General Electric decided to develop a market for the surplus GE 750's. The fifteenth locomotive to come down the assembly line was completed as a 4'8½" gauge machine and offered to the Milwaukee as a demonstrator.

GE750 on the General Electric test track at Erie, Pennsylvania./Photo courtesy Milwaukee Wisconsin Public Library.

The timing of this offer worked out well because of an unexpected turn of events within the Milwaukee's management. Diesel advocate J.P. Kiley became the Milwaukee's Operating Vice President and appointed Laurence Wylie as the new head of the Electrification Department.

Wylie accepted the job, but quickly and quietly worked to preserve the electrification which he saw as superior to diesels. Among the problems he faced was how to prove this superiority when all of the electrics on hand were well worn and obsolete. He welcomed the opportunity to test one of the new machines, and in early 1949 the first Little Joe arrived on a three month loan. It was painted olive green with a white stripe and it bore the number GE-750.

Tests were conducted on both the Coast and the Rocky Mountain Divisions. Performance was monitored by observers from the Electrification Department, the Mechanical Department and by representatives from G.E.

> "I was on the test run of the GE-750 when it was tried out across the system," says Barry Kirk. "The tests were not too successful. That contributed to the Milwaukee's decision not to buy the Little Joes until a couple of years later."
>
> "The biggest problem was that the Little Joe couldn't handle the tonnage that the engineers from G.E. said it would. The G.E. people had based all of their calculations purely on horsepower and hadn't taken tractive effort into account. As you know, a four-wheel drive jeep will go lots of places that a high horsepower sedan won't. This is because the jeep can get traction. A locomotive's pulling power must be measured in terms of tractive effort as well as horsepower."
>
> "Though a Little Joe could pull a train faster than a two-unit set of Freight Motors, the Freight Motors had a lot more weight on their drivers. The Freight Motors could start a train the Joe would slip its drivers with and they could outpull a Joe at 15 to 18 mph."

Seattle based electrical engineering consultant Laurence Wylie at age 71 in 1963./Photo from the Donald Wylie Collection.

"Other problems that came up were with the pantographs. The tests were conducted in the middle of winter and G.E. had used some very constricted air piping for the lines to the pantograph air cylinders. The pipes kept freezing up and we weren't able to raise or lower the pantographs."

"The original pantographs that came on the demonstrator were a General Electric design that differed from those used on the Milwaukee. They didn't work well and they really had to stretch out to reach our 24-foot high wire. When they stretched out like that, they tended to whip around badly. We replaced those pantographs with some standard Milwaukee ones. When we installed them, we put them on raised platforms so they wouldn't have to stretch so far in order to reach the wire. All in all, the tests went poorly. When they were over, some representatives from the Mechanical Department announced that the new locomotive was less valuable than a GP7. Their announcement was really no surprise though. They were against electrification and electric locomotives."

At the conclusion of the tests, the GE 750 was returned to G.E. and the Milwaukee's official position was non-interest. Regardless of what others thought, Laurence Wylie was impressed by the potential of the new locomotive. As a result, he made a trip to Erie, Pennsylvania a few months later to look over the locomotives and make an offer.

"I negotiated all day, all night and all the next day to hammer out a deal," said Wylie. "I told them that those Little Joes would be impossible to sell and had no value other than scrap. Then I offered them the scrap metal price of $1 million for all 20 Little Joes 'as is' and all the spare parts they had on hand. I convinced them that it was in their best interests to have those motors running rather than just cut up for scrap. The representatives from G.E. agreed to it, and we had a deal until Kiley vetoed it. Kiley didn't know much about electrification and didn't like it. He was planning to dieselize everything and he didn't want to spend any money on new electrics."

Unpredictable international politics affected the destiny of the Little Joes a second time in the summer of 1950. On June 25, North Korea invaded South Korea. U.S. military involvement followed shortly thereafter and war related traffic surged on the Milwaukee. As a result, the Milwaukee was power short and there seemed no immediate way out. Not all of the diesels on order had arrived, and many of the Milwaukee's steam engines had been stored unserviceable. As the pressure to move traffic mounted, Kiley gave Laurence Wylie authorization to purchase the surplus electric locomotives. Unfortunately, eight of the fleet and all of the spare parts had already been sold. What is more, G.E. was now in the negotiating driver's seat. The price demanded was $1 million for the remaining twelve on an "as is" basis.

"I tried to buy the other locomotives from the South Shore and from the Brazilians," said Wylie, "but no one would sell."

The South Shore purchased three shortly after the original Milwaukee deal fell through. With technical guidance from G.E., the South Shore modified them to run beneath the 1,500 volt DC trolley wires of its Chicago area freight hauling suburban commuter line.

The 3,000 volt Cia Paulista Railway bought five to run its 5'3" gauge electrification based in Sao Paulo, Brazil.

The 12 purchased by the Milwaukee were taken to the Milwaukee, Wisconsin, shops to receive the modifications which were necessary before they could be placed in service. All needed to be painted in Milwaukee colors, nine needed conversion from 5' gauge to 4'8½" gauge, and several required the removal of buffers and Russian couplers. Two were to receive additional changes to provide for their use as passenger locomotives. All modifications were completed in the four months from August through December of 1950. As work was completed, the Joes were towed to Deer Lodge, where they were tested and placed in service.

The Joes were 88'10" long over couplers, 14'5" high over cabs and their width was 10'7". They rode on 47½" drivers and the idler wheels in the pilot trucks had a 37" diameter. All wheels were spoked with steel tires on cast steel centers. The total wheel base was 77'10" and the rigid wheel base was 20'7". This rigid wheel base was longer than that of any other Milwaukee Road steam, diesel or electric locomotive. Each Joe had eight 1,650 volt DC traction motors which were wired in pairs in order to accept the designed trolley voltage. The motors were geared to the axles in a ratio of roughly 4 to 1. The locomotive chassis consisted of two heavy cast steel driving trucks linked by an articulation joint. The pilot trucks were mounted under the ends of each chassis half.

The 77' streamlined superstructures contained control cabs at each end. Behind the cab walls were two walkways linking the cabs and between the walkways sat the control equipment. This included resistor banks, traction motor blowers, air compressors, contactors, a high speed J-R type circuit breaker and an M-G set which produced current for control equipment and train lighting. A door in the front of each cab led to the nose. Inside the noses were hand brake wheels and access to the headlights and number board lights. A door in the nose allowed crews to enter and exit the locomotives through the noses. On the chassis, in front of the noses were tall M-U jumper posts.

Top:
Blue and white FEPASA 6454 sits in the yard at Sao Paulo Brazil. The engineer's name is proudly displayed on the side. As a result of a merger, the FEPASA succeeded the Cia Paulista./Photo courtesy FEPASA

Middle:
Looking every bit a close relative of the Little Joes, Great Northern Class W-1 5019 sits at Skykomish, Washington 11-5-53./Photo by Casey Adams.

Right:
Norfolk & Western #222, formerly Virginian #126, hauling coal on the old Virginian mainline in 1960. The EL-2B's were powerful and rode like Pullmans./Photo courtesy Norfolk and Western Railway.

Orange and maroon South Shore 803 at Michigan City, Indiana. 2-10-77. The South Shore called these motors "The Eight-hundreds" rather than Little Joes./Photo by Don Kaplan.

STATISTICAL COMPARISON OF THE LITTLE JOE AND OTHER ELECTRICS OF ITS ERA

	Great Northern Class W-1	Virginian Ry. Class EL-2B	Milwaukee Road Class EP-4/EF-4	SZD (Soviet Railway) Class VL8
Year Built	1947	1948	1948-1949	1953-1967
Number Owned	2	4	12	1,500
Wheel Arrangement	B-D+D-B	B-B+B-B+ B-B+B-B	2-D+D-2	B+B+B+B
Traction Motors	12	16	8	8
Superstructure	one unit	two unit	one unit	two unit
Height Over Cab	15'6"	15'6"	14'5"	not available
Total Length	101'0"	150'8"	88'10"	90'3½"
Total Wheel Base	85'9"	133'8"	77'10"	not available
Rigid Wheel Base	16'9"	9'0"	20'7"	approx. 10'
Weight in Pounds	720,000	1,000,000	586,600	368,000
Weight On Drivers	720,000	1,000,000	443,000	368,000
TE at 25%	180,000	250,000	110,000	92,000
Continuous TE	122,000	162,000	75,700	66,800
Continuous HP	5,150	6,800	5,550	5,712
Max. Safe Speed	67.5 mph	50 mph	68 mph	62 mph
Current	11,000 vac	11,000 vac	3,300 vdc	3,000 vdc
ROUTE DATA				
Electrified Route	72 miles	134 miles	654 miles	834 miles*
Ruling Grade	2.2%	2.07%	2.2%	1.5%
Maximum Curvature	10 degrees	12 degrees	10 degrees	5 degrees

*SZD electrified route mileage listed is for 1957. It includes the 700-mile Trans-Siberian, electrification which was completed in 1955. The SZD's 3,000 volt mainline network grew to approximately 12,000 miles by 1969. Suburban lines and 25 KV electrification is in addition to that total.

Data compiled by the author.

The Joes as built weighed 273 tons each. This was increased to 289 tons on E20 and E21 when train heat boilers and water tanks were installed. The installation of concrete slabs below the pantographs of E70 through E79 brought their weight to 293 tons.

When General Electric delivered the Little Joes in 1950, the specifications indicated they would produce 4,750 horsepower continuously and 5,120 for up to one hour. Testing on the Milwaukee showed they would produce 5,110 hp continuously and 5,530 hp for one hour, when operating on 3,000 volts. If operated at 3,300 volts, they produced 5,550 hp continuously and 5,860 hp for up to one hour.

The Joes were good locomotives and were well received on the Milwaukee but their design was not ideal for a railroad with 10 degree curves, mountain grades and heavy trains. With only 443,000 pounds on the drivers, their tonnage rating was 90% of that assigned to the two-unit G.E. Freight Motor sets. Additionally, the 20'7'' rigid wheel base produced heavy wheel flange wear on sharp curves. If the Milwaukee had ordered a group of locomotives designed specifically for its route, the result might have been direct current locomotives resembling the Virginian's 11,000 volt AC, EL-2B's. Those million pound, two-unit locomotives were 150' long, had all axles powered and rode on eight 2-axle trucks. Because of the use of 2-axle trucks, the rigid wheel base was only 9'. Thus the EL-2B's did not produce heavy flange and rail wear. In 1947, General Electric and Westinghouse jointly proposed building 35 such locomotives for the Denver and Rio Grande Western. Although the Grande chose not to electrify, and thus did not buy them, the Virginian bought four.

In 1947, the Great Northern also bought giant new electrics. Theirs had the same wheel arrangement as the Joes, but all axles were powered and the locomotives weighed 720,000 pounds. These 101 foot locomotives only had a 16'9'' rigid wheel base, but they produced noticeably more rail wear than did diesels.

Although General Electric looked upon the Joes as locomotives for export and did not feature them in G.E. advertisements, many people on the Milwaukee were impressed with them. They were seen as a logical choice for replacing all of the older motors on both electrified divisions. Several times Laurence Wylie attempted to negotiate deals whereby G.E. would construct a group of 10 to 25 more, but the deals never quite came together. The asking price was reportedly $650,000 each. This was roughly equal to the price of a four-unit set of F7 diesels and was more than the diesel oriented Milwaukee management was willing to pay.

With only 12 Joes on a 654-mile electrification, the Milwaukee assigned them all to the Rocky Mountain Division. The Rocky Mountain Division had more traffic than the Coast and this made for better locomotive utilization. Additionally, the Rocky Mountain Division had long stretches of track where high speed running was practical. The Joes could easily roll freight trains at 50 mph to 70 mph. This made for dramatic improvements in Rocky Mountain Division operation since such speeds were beyond the capabilities of the G.E. Freight Motors. On the Coast Division, few places existed where speeds over 50 mph could be maintained for significant distances.

In the late 1950's, the Milwaukee once tested the idea of hauling a train, Joes and all, across The Gap with diesels and then running it on the Coast Division. This, however, proved to be of little value and it interfered with the assignment rotation of Coast Division locomotives. During most of their few appearances on the Coast Division, the Joes were simply towed to and from the Tacoma shops dead in the train. Coast Division engineers did not have training in how to run them properly. Additionally, Coast Division substation operators had no experience feeding current to locomotives which could draw so much current so rapidly.

A 6750-HP ELECTRIC LOCOMOTIVE WITH A MILLION POUNDS ON DRIVERS

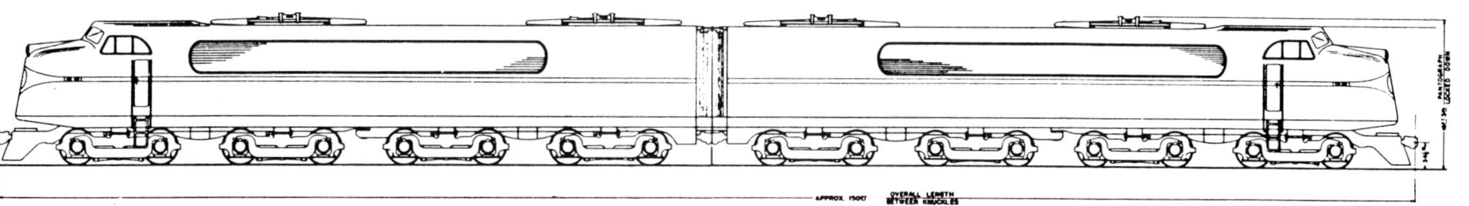

PROPOSED LOCOMOTIVE FOR THE DENVER & RIO GRANDE WESTERN RY.

PRINCIPAL DATA AND DIMENSIONS

Wheel Arrangement	2[(B-B)+(B-B)]
Total Weight	1,000,000 lb
Weight on Drivers	1,000,000 lb
Number of Driving Axles	16
Weight per Driving Axle	62,500 lb
Continuous Rating	
At Rail	6,750 hp
Tractive Effort	150,000 lb
Speed	16.9 mph
Adhesion	15.0 per cent

Starting Tractive Effort	
25% Adhesion	250,000 lb
Maximum Operating Speed	75 mph
Number of Traction Motors	16
Line Voltage	12,000
Phases	Single
Length Overall, between Knuckles, Approx.	150 ft
Width Overall	11 ft 0 in.
Height over Pantograph, Locked Down	16 ft 0 in.

THE PASSENGER JOES
Class EP-4

The two Little Joes which were selected to operate in passenger service comprised the EP-4 class. They were the last two Joes to be regauged and repainted by the Milwaukee, Wisconsin, shops. When the work was completed, they were assigned the numbers E20 and E21. This was in keeping with the company policy of assigning low numbers to passenger motors.

In order to adapt them to passenger service, the E20 and E21 received additional modifications which the other Joes did not. These were roller bearings and train heat boilers.

Roller bearings were installed in order to avoid bearing failures during sustained high speed operation. They were applied to all axles and to the traction motor armatures. The result was two locomotives which rolled so freely that crews were warned to set the brakes or block the wheels if leaving the locomotives unattended in a strong wind or on a slight grade. The roller bearings so successfully resisted failure that they were later installed on freight Joes as replacements for plain brass bearings that failed.

In order to provide space for the train heat boilers, control equipment was removed from the cab on the "B" end of each locomotive. In its place the shop crews installed semi-automatic, high capacity Elesco steam generators. Although Clarkson and Vapor brand steam generators were standard on Milwaukee diesels, the Elesco units were chosen for the Joes because they were lighter and more compact. This made them easier to install on locomotives which had not been designed to carry them.

As a back-up in case of boiler failure or the need to use a freight Joe in passenger service, the Milwaukee converted two old steam locomotive tenders into boiler cars. Whenever they were needed, they were coupled behind the locomotive.

The E20 entered service on December 21, 1950 and the E21 followed on January 2, 1951. They were assigned to the Rocky Mountain Division to supplement the 7 remaining Westinghouse Motors.

Passenger Joe E21 at Harlowton with a boiler car in tow./Photo by Richard Steinheimer.

E20 and E21 were unarguably the Milwaukee's best passenger motors and quickly became favorites of the train crews. They were well suited to passenger service because of their quick acceleration and high speed capabilities. Additionally, each could easily cross the Rocky Mountain Division with 20+ passenger cars. This was power roughly equal to a three-unit set of Fairbanks-Morse diesels and it exceeded the capacity of single Westinghouse Motors by roughly 40%.

Swede Hansen, who had experience with the passenger Joes both as an engineer and a fireman, compares them with their predecessors, the Westinghouse Motors:

> "The Joes were a speedier engine and newer, so you got more out of them. Everyone who worked on them liked them better than a Westinghouse."

Retired engineer Bill Lintz agrees, adding:

> "Those Joes were the most beautiful engines you could imagine. They got heavy use but they just stood up and kept running all the time. E20 and E21 ran 400 miles a day, every day, making the trip between Harlow and Avery, and they hardly ever needed to go into the shop."

Coast Division Road Foreman of Engines, King Clover, ran them while working on the Rocky Mountain Division and recalls being impressed:

> "The first time I ran one, the regular engineer told me I was too conservative with the throttle. He said to just open it up. When I did, our train took off like a rocket. You could really wallop a passenger train with a Joe. They had super acceleration because you could build the amperage up much higher than with any of our other motors. The only limitation was that if you got it too high, they would slip their drivers."

"The speedometer went up to 85 mph and sometimes those guys would have the needle banging off the peg. On the Rocky Mountain Division they really ran fast. I was used to that when I ran steam in Europe during World War II and after the war when I was an engineer on the Southern. On the Milwaukee's Coast Division though, 60 to 65 mph was about as fast as we went."

When the Joes entered service, there were two trains a day each way on the Rocky Mountain Division; the streamlined Olympian Hiawatha and the mail carrying Columbian, which made the local stops. All were pulled by electric locomotives.

Although the Joes were very good passenger power, plans to remove them from passenger service were developed after only a year of operation. When Laurence Wylie's initial efforts to purchase additional Joes failed, he decided to concentrate the modern electric motive power in the area where it would do the most good. That was Rocky Mountain Division freight service. Other passenger motors could meet the passenger schedules, but only the dual service Joes could be used to improve the Milwaukee's schedules for freight.

The change would have been implemented quickly, but replacing the EP-4's proved to be more difficult than first expected. The ranks of the Westinghouse Motors dwindled rapidly as they came up for major repairs and were scrapped. The rebuilt G.E. Motors, E22 and E23, had less power than expected and their plain brass bearings tended to run hot in sustained high speed operation. The legendary Bipolars, which had also been rebuilt, proved to be unreliable for a number of reasons. Thus E20 and E21 continued to operate in passenger service until 1958 when the Rocky Mountain Division's dwindling passenger traffic was turned over to diesels and Bipolars.

Boilers were removed from the E20 and E21 when they entered the freight locomotive pool, but they retained their low passenger numbers through all their years in service. They operated until 1974 when the electrification was shut down and replaced by diesels.

The passenger trains they once hauled were not so long lived. In 1955, the Columbian ceased to operate in electrified territory when it was cut back to Aberdeen, South Dakota. In 1961, the Olympian Hiawatha ceased running to Tacoma and its famous name disappeared from the timetables. It was first cut back to Deer Lodge and then in 1964 the remaining route was cut back to Aberdeen.

E20 speeds westbound at Bearmouth, Montana with train #15 on June 21, 1955./Photo by Bruce Black.

*On a cold day in March 1953, the in-bound Butte Anaconda and Pacific mixed train meets E20 and the westbound Olympian Hiawatha at the old Butte depot. For the benefit of the BA&P, both roads run on 2400 volts in the depot tracks./*Photo by Richard Steinheimer.

*The Olympian Hiawatha's distinctive observation lounge as train #16 prepares to leave Three Forks, Montana 5-29-50./*Photo by John C. Illman.

Inside the Sky Top Lounge between Harlowton and Miles City, 5-29-50./Photo by John C. Illman.

THE FREIGHT JOES
Class EF-4

This class originally consisted of ten locomotives numbered E70 through E79. In 1958, the group increased to 12 as E20 and E21 were placed in freight service.

During 24 years of service on the Milwaukee, no major modifications were made on this group.

In the early 1950's, the MU jumper post was removed from the "A" end of each Little Joe. It was felt that the posts were unnecessary since operating Joes in groups of more than two would overload the electrical system. In addition, it was the Milwaukee's practice to run pairs coupled together at the "B" end.

In the late 1950's, the need to move more tonnage per train in electrified territory became a pressing concern. Adding more electric locomotives to each train was out of the question. The operation of diesels with electrics was impractical because differences in the control systems prevented MU operation. Any diesels used would require separate train crews. In order to solve this problem and expand the utility of electrics, Laurence Wylie developed a low cost MU controller. It was applied to the Little Joes in 1958 and 1959. Miniature diesel throttles were installed next to the electric throttles. The two were linked by a removable lever which moved the diesel throttle in proportion to that of the electric. The diesel throttle was con-

E21 and E76 in Deer Lodge on August 17, 1960. Note the roller bearing journal covers and the blanked out rear cab windows which differentiate E21 and E20 from other Joes./Photo by Doug Cummings.

E74 at Deer Lodge, July 17, 1960./Photo by Doug Cummings.

74 at Deer odge June 21, 955./Photo by ruce Black.

E73 and E72 at Deer Lodge 5-27-55./Photo by Casey Adams.

The engineer on E75 prepares to head east from Three Forks to Harlowton in the summer of 1972./Photo by Noel T. Holley.

nected to the trailing diesels through normal diesel MU wiring and hoses.

Through the use of diesel "Boosters," tonnage capacity in the electric zones was increased. Standard practice on the Rocky Mountain Division was to run one GP9 behind a pair of Little Joes. In the late 1960's, when it became common to run groups of diesels straight through from Chicago to Tacoma, single Joes were often used ahead of several diesels.

According to retired engineer Swede Hansen:

> "Diesels alone on the mountain had problems because their transition controls were completely automatic. If you couldn't get going fast enough to go through transition, they would just poke along. The best way to run trains was when they put one Joe on the point, ahead of three or four diesels. The Joe would keep the diesels wound up so that you could really move."

Beginning in 1964, more extensive modifications were made on the fleet. Joes E70 through E79 became single cab locomotives, as control equipment was removed from cabs on the "B" end. These rear cabs had seldom been used and the ex-passenger Joes had always run without them. The primary reason for removing the "B" end controls was to create a new place for the J-R breakers. General Electric had installed these high speed electro-magnetic circuit breakers in a cabinet near the contactors and this proved to be a bad location. According to

Storm clouds blow over the Tobacco Root Mountains southeast of Vendome as #263 grinds up the 2% on the east side of Butte Hill. Joes E76/E73 and GP40-2's 19 and 17 leaning into the crossing of the Continental Divide with 106 cars on the drawbar. 5/10/74./Photo by Ted Benson.

Little Joe E77 leads a six-unit mix of GE power (U25B's/U28B's/U30B's) out of Avery on #264. The GE's aren't common Rocky Mountain Division power... neither are Joes anymore! May, 1974./Photo by Ted Benson.

Train #266, a low priority "dead freight." E79 and E78 are the power. The failed E76 was added to the 75 cars at Alberton in the summer of 1972./ Photo by Ted Benson.

*It's mid-May, 1974, but winter has yet to depart the Bitterroots as #263 charges uphill out of Tarkio behind E20/E77—all four pantographs are up to cut the ice on the overhead./*Photo by Ted Benson.

retired Electrification Department head Barry Kirk, the ionized gasses created by the operation of the contactors caused the J-R breakers to operate erratically and burn up. Moving them away from the other electrical equipment solved the problem.

Other modifications to the fleet included the installation of sealed beam headlights and extremely rugged snow plows on the "A" end of each Joe. The shops also began repainting Joes in the orange and black color scheme which had been used on Milwaukee Road freight diesels for over ten years.

With only one exception, the fleet was maintained to a fairly uniform appearance. That one exception was E78, which was heavily damaged in a 1966 derailment. It was rebuilt at the Milwaukee, Wisconsin shops. In order to simplify this job, the shop utilized some spare parts which had been manufactured for EMD carbody diesels. These were cab window sections and chrome plate side grills. The E72 received damage of equal severity in the same wreck, but it was restored to its previous appearance. The Deer Lodge shops performed that work.

Throughout their 24 years of operation, the power and reliability of the Joes made them legends. The G.E. Motors were slow and due to age, they were rapidly wearing out. Diesels could match the Joes for power only if operated in groups and even then they could not match the Joes for speed. The tonnage ratings used by dispatchers were calculated based on

Little Joes E75 and E70 passing through Butte westbound with train #263 4-1-62./Photo by R.V. Nixon.

A power swap at Haugan with #263 brought the herd of GE diesels on #264 back to Avery later in afternoon of May 13, 1974—here Joes E21/E78 lead the lashup of 5503 / 6005 / 5006 / 6002 / 5003 / 5002 over the Kelly Creek Bridge above Adair with 91 cars, 3267 tons trailing./Photo by Ted Benson.

June 13, 1974, train #264 eastbound through Drexel, Montana with E72, 4005, 2058, 16, 4008./Photo by Dick Dorn.

The rebuilt E78 at Avery, Idaho June 24, 1971. The "B" end has been blanked out. Steel plates replaced the windows and number board glass. The "B" end pilot has also been removed./Photo by Jack Powers.

minimum continuous operating speeds. For diesels, this was 18 mph. For Joes, it was 24 to 28 mph. In addition, the diesels experienced broken valves and generally increasing mechanical problems if operated at maximum power for extended periods of time. Regardless of performance demanded, the Joes were reliable day in and day out. What is more, they could exceed their design specifications whenever the need arose.

Barry Kirk, who headed the electrification from 1963 to 1971, recalls an incident of the type which made the Little Joes a favorite of train crews who often struggled to get heavy trains over the line.

> "The Little Joes could really put out some power," says Kirk. "One time I was riding over the Butte Hill on a train with two Little Joes and some diesels. The diesels broke down while we were on the grade so the engineer stopped the train. We didn't want to double the hill, so I gave the engineer approval to see if the Joes could make it by themselves. It was a good day with dry rail and no wind to blow sand off

E76 has failed and was left behind. E75 is going it alone running blind end first at 55 mph with "dead freight" #265. On this summer day in 1972, she is west of Superior and headed for the wye at St. Regis. There she will be turned./Photo by Ted Benson.

E78 suffered major damage to its body and electrical equipment in the 1966 wreck which also wrecked E72. The Milwaukee, Wisconsin shops rebuilt it with EMD diesel F unit parts. These include cab sections, louvers and chrome plated side grills. 9-13-73./Photo by Bruce Black.

E72 by the Harlowton roundhouse on July 3, 1973. It suffered major body damage and a bent frame in the 1966 wreck. The Deer Lodge shops restored it beautifully to the original appearance./Photo by Bruce Black.

On the night of the last electric operation, the Deer Lodge roundhouse plays host to E72, E20 and E79. June 15, 1974./Photo by Dick Dorn.

> the track. Also, we were right beside the Janney substation when we started, so I knew we could get full voltage. Running at 475 amps per traction motor, the Joes pulled the train over the hill with no problem at all. After the trip, I did some calculations and they showed that the Joes had to produce 6,700 to 7,000 horsepower each at 30% adhesion to pull that train."

Although most aspects of the Little Joe's performance drew praise, there were two which were the center of some criticism. These were the large amount of current the Little Joes drew, and their tendency toward wheel slips.

The Little Joes' appetite for current was so great that two of them could exceed the capacity of the power supply system if worked hard. Based on their one hour rating at 3,000 volts, each Joe could draw 3,000 amps. The rule of thumb maximum current available for a single train was 4,000 amps. According to a Rocky Mountain Division Trolley Foreman, pairs of Little Joes sometimes caused the copper trolley wires in the Belt Mountains to get so hot they stretched and sagged. This created trolley tensioning problems until heavy aluminum feeder wires were installed in order to deliver more current. This did not, however, solve the problems of substation operators. They had to be prepared to lower voltage while shifting and sharing the load, in order to protect their equipment.

The Little Joes had a reputation for being slippery, but it was largely undeserved according to Barry Kirk:

> "Most of the wheel slip problems were due to improper operation of the locomotive. The wheels would slip if the traction motors were drawing too many amps. That was the Little Joes' safety valve."

Occasionally, in the mountains, there were slippage problems which were not the result of engineers trying to get too much out of their locomotives. The third and sixth sets of drivers sometimes slipped on mountain grades while the Joes were pulling trains well within their limits. According to Kirk, close observation showed that the Joes' long frame could not track

Train watching in the mountains was at its finest in the canyon above Avery. Here a heavy freight with a "Double Joe" and three diesels ascends the 1.7% grade to St. Paul Pass. 8-20-70./Photo by Jim Fredrickson.

June 10, 1974, Little Joe E77 arrives at Two Dot, Montana trailing smoke from journal #4. She is set out and train #264 struggles on without her. Later, she will be towed dead to Deer Lodge./Photo by Dick Dorn.

properly on tight curves. As a result, the wheel tread on the third and sixth sets of drivers was only partially on the railhead, making them prone to slippage.

The need to replace the G.E. Motors was an issue when the Little Joes were purchased in 1950. As time went by, the Milwaukee developed a greater and greater need for a new electric fleet. Electrification Department studies predicted that the last of the GE Motors would be gone by 1976 and that the Joes would need replacement by 1983. The GE Motors wore out more quickly than expected. They were retired in droves during the 1960's and the 12 Little Joes became the only true road locomotives using the 654 miles of mainline trolley. While the Milwaukee pondered what to do with its electrification, the void was filled by diesels.

Laurence Wylie had retired in 1956 but continued to serve the railroad as a consultant. His recommendations on motive power were that the Milwaukee should seek locomotives as rugged and well designed as the Little Joes. He expressed concern, however, that such locomotives may be unavailable in the U.S. General Motors had no experience constructing electrics and he felt that General Electric's work since the early 1950's showed a lack of innovation and quality. In addition, G.E. had few remaining experts on 3,000 volt electrification. The men who designed the Joes were Wylie's age and they too had retired.

As part of a 1968 report, Wylie prepared specifications and sketches of two types of locomotives he felt would meet the needs of the railroad. One had a body resembling a Little Joe, but rode on C-C trucks. It was rated at 4,680 horsepower. The other resembled two carbody type diesel units back to back. It was a B-B + B-B locomotive rated at 6,240 horsepower.

According to Barry Kirk, proposals were received from General Electric and General Motors in 1968 and 1969. G.E. offered to build 5,000 horsepower C-C straight electrics for approximately $500,000 each. The appearance of the locomotives was not specified. As a lower cost alternative, G.E. also proposed 4,000 horsepower motor-generator electrics for $400,000 each. This M-G type would resemble a U-36C with a 3,000 volt DC motor driving the 600 volt DC traction generator. An offer to build M-G electrics was also received from General Motors, but Kirk was not interested in M-G electrics. He felt their higher maintenance costs and lower efficiency would make them a more expensive choice in the long run.

Most real advances in electric locomotive design were being made in Europe. Electrics were the dominant type of motive power there and several builders had recent experience producing new 3,000 volt designs. The Milwaukee inquired about having Brown-Boveri build locomotives, but it appeared this company could not build anything soon enough.

At night on the eve of the last day, Little Joes E73 and E20 await the call to roll the last electric powered train./Photo by Dick Dorn.

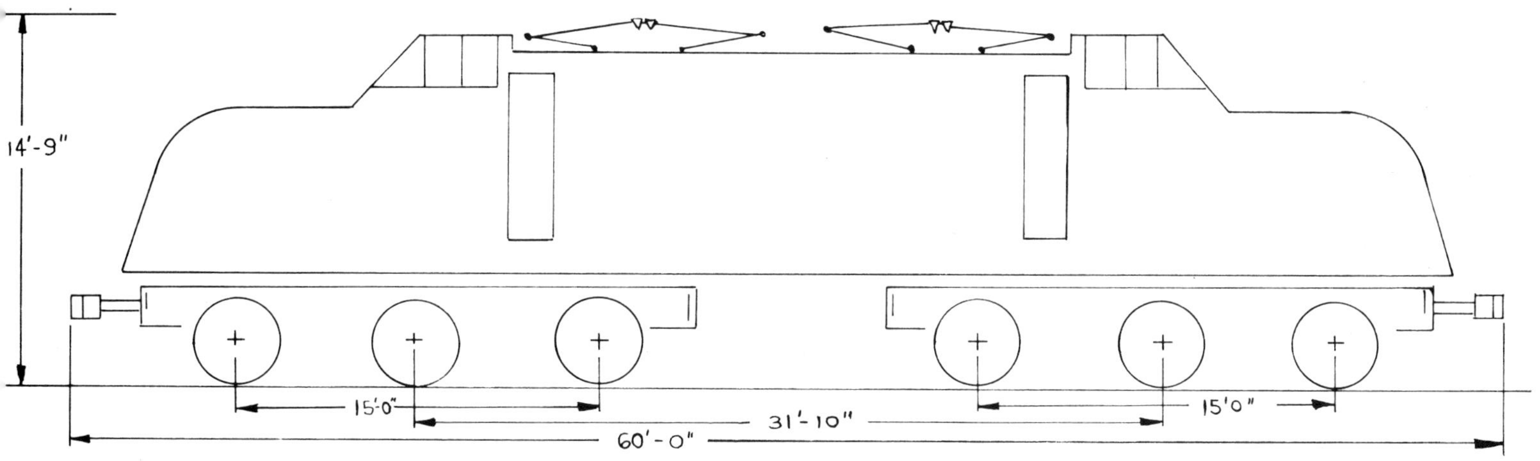

Skeleton outline of 4680 hp, 3300 volt direct current locomotive.

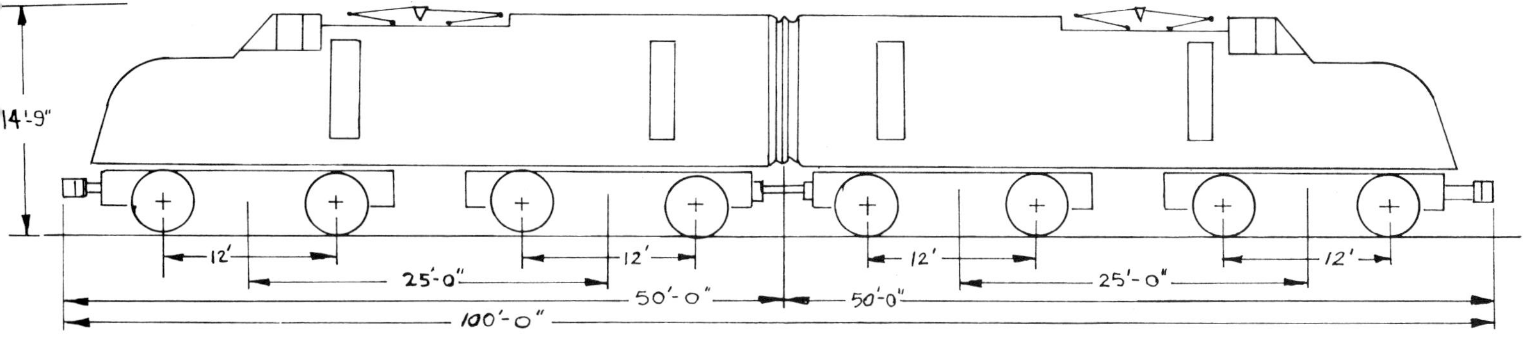

Skeleton outline of 6500 hp, 3000 volt direct current locomotive. *Both Drawings by Laurence Wylie.*

E73 and E20 have arrived in Deer Lodge with the final electric run, June 15, 1974. Engineer Bill Lintz, Conductor Ed Johnson and Brakemen Daryl Arfstrom and Dewayne Eisenbarth pose near the depot./Photo by Dick Dorn.

No decisions were made regarding the Milwaukee's western locomotive fleet until 1973. Following a review of company economics, the Milwaukee's management concluded the company could not afford to revitalize its electrification. Diesels became successors to the electric fleet.

The death knell rang for the Little Joes when October 31, 1973 was set as the final day of electric operation. As it approached, the Joes were removed from service one by one and stored behind the Deer Lodge shop. Dieselization was essentially complete when changing world affairs changed the fate of the Joes for one last time. On October 17th, the Arab Oil Embargo began. Fuel shortages gripped the land, prices rose to unprecedented levels, and the Joes kept rolling as the deadline passed. The Joes which had been stored were returned to service. Electrification continued, but the die was cast and the end came eight months later. June 15, 1974 was the final day and the team of E73 and E20 brought the last electric powered train into Deer Lodge.

Many on the Rocky Mountain Division regretted the loss of the Joes, and their opinions are summed up by the words of retired engineer Swede Hansen:

> "There ain't a diesel made that will compare with a Joe. A damn diesel is just like an automobile, there are too many things that can go wrong with them. They just can't take it. The diesels were nice when they were new, but they went to pot after a few years. I ran everything the Milwaukee had in my day and the Joes were the best engines the Milwaukee ever owned."

Following retirement, the Joes were stripped of reusable parts such as traction motors. All but one were then sold to scrappers in Seattle and Chehalis, Washington. E70, the one remaining Joe, was donated to the city of Deer Lodge. It is displayed on a short piece of track near the courthouse. At the time it was donated, E70 was painted orange and black. It is the ex-demonstrator locomotive GE-750.

Two similar locomotives which did not run on the Milwaukee are also still in existence. South Shore 802 is part of the Baltimore and Ohio Transportation Museum collection in Baltimore, Maryland. The 803 is in the Illinois Railroad Museum in Union, Illinois.

Little Joe E77 and four GP-40's at Avery, Idaho following a snow storm./Photo courtesy Milwakee Road News Bureau Collection.

On June 11, 1974, Joe E74 waits patiently outside the Harlowton roundhouse. Her next trip to Deer Lodge will be her last./Photo by Dick Dorn.

THE ROTARIES

On the Milwaukee's Pacific Extension there were many areas where the railroad battled with snow. The onset of winter typically brought a succession of snow storms, some of which had the potential to tie up the line.

In most areas, the storms brought light dry snow which would drift in the wind, and accumulate only in cuts. In these areas, locomotive pilot plows and wedge plows pushed by locomotives were usually enough to keep the tracks clear. There were, however, two areas which often required rotary snow plows. Snoqualmie Pass, in the Cascades and St. Paul Pass, in the Bitterroots had snow which tended to be moist and heavy. It fell in deep blankets which did not drift, and if left unplowed, it could accumulate to depths of 15 feet. New snow was most often measured in inches, but major storms could dump a foot an hour in certain places.

Sliding snow was another source of problems. Most of the Milwaukee trackage in the Cascades and Bitterroots was built on the sides of mountains. It could be blocked by massive slides if winter thaws and rain caused the snow above the rail line to break loose. Removal problems were compounded by the fact that slides generally contained trees and rocks pulled down by the snow.

In order to protect the line, the Milwaukee owned two types of rotaries. Standard rotaries

X900212 at Tacoma, Washington roundhouse shortly after being rebuilt as an electric plow. Date unknown./Photo by Laurence Wiley.

Rotary X900207 was rebuilt at Milwaukee, Wisconsin. Its diesel traction motors could not operate from the 3000 volt trolley. Shown at Tacoma, Washington 2-80./Photo by Jim Fredrickson.

A few years after her conversion to electric power the X900215 was painted maintenance-of-way yellow. She spent the rest of her career that color and was generally found near this spot in Tacoma awaiting snow. The photo is dated 7-25-59./Photo by Doug Cummings.

The second electric plow, X900215, shown near the Tacoma roundhouse. Her traction motors came from a Westinghouse Motor. Date of photo unknown./Photo by Laurence Wylie.

were used whenever the snowfall was too heavy for wedge plows. If the problem was slides, then "Slide Rotaries" were called instead. The Slide Rotaries had fewer blades on their plow wheels, but the blades were more ruggedly built in order to effectively chew up trees and rocks.

Rotaries were most often used on the 39 miles of line from Cedar Falls to Easton, on the Coast Division, and the 57 miles of line from Avery to St. Regis on the Rocky Mountain Division. The rotaries were based in Tacoma, Avery and Deer Lodge. Until steam was retired, each division generally had three rotaries.

The steam rotaries contained an oil fired boiler and a pair of cylinders linked to the plow wheel through rods, cranks and gears. The engineer and the wheel operator rode in the front while the fireman rode in the rear by the firebox.

> "It was wet and miserable on the old steam plows from the time you got on until the time you got off," says Ralph Edwards. "You had to wear a long coat, rubber boots and a rubber hat to try to stay dry. Even then you were wet because you had snow blowing in through the open window in front of you, and you were sweating from the heat of the boiler behind you. Those plows were a fireman's nightmare because they didn't have a big enough boiler for the size of the cylinders. A fireman was fighting for steam and water the whole way if you got into heavy plowing. You had no way of communicating with him, so he would just have to listen to your exhaust. If he didn't open up the firing valve when he heard you open up the throttle, you might run out of steam. Quite frequently you would run out of water if you were in-

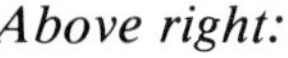
Above left:
Inside X900212. The large lever equipped console is the 3000 volt controller. The smaller one next to it is a 600 volt controller for diesel supplied current./Photo by Laurence Wylie.

Above right:
Cab of plow X900215./Photo by Laurence Wylie.

Right:
Next to the aisle is one of the big 3000 volt traction motors which turn the plow wheel on X900212./Photo by Laurence Wylie.

to heavy plowing and then you would have to run for a water tank. The plows were a miserable job, but everyone wanted to work them because the pay was so high and it was a 16-hour shift.''

''In the winter of 1949, the snow was pretty bad. It snowed that winter, and then it started to thaw. It froze up, and then snowed again. It thawed right after that and the whole mountain came down in slides. We had one slide come down and block the west end of Snoqualmie Tunnel. Everything was tied up at Hyak including the rotary. They ran the rotary through the tunnel from the east and they would plow as long as they could stand the heat and smoke. Then they would back out and wait for the smoke to clear before they went in again. It took them a long time to get the tunnel open.''

Nineteen fifty-four was the beginning of the end for steam rotaries on the Milwaukee. Laurence Wylie drew up plans for converting the 900212 to electric power and the Tacoma shops performed the work. The boiler and drive gear in the 212 were replaced by two big GE traction motors. A new steel body was built and on its roof was a pantograph for 3,000 volt

Hyak, Washington, after a heavy snow. A steam rotary sits near the substation. In the foreground are the lodge and passenger sheds for the Milwaukee Ski Bowl. Date unknown./Photo by Laurence Wylie.

Steam rotary X900212 plowing at East Portal, Montana./Date and photographer unknown.

GP-9 connected to X900215 to provide 600 volt current to the motors./Photo by Laurence Wylie.

Summer 1972 finds rotaries 900212 and 900214 inside the Avery roundhouse./ Photo by Noel T. Holley.

operation. On the rear were power cables for 600 volt supply from trailing diesels.

The plow was tremendously powerful when operating on 3,000 volts, but the plow wheel turned far faster than the optimum plowing speed of 150 rpm. It could be slowed by heavily exciting the traction motor field coils, however, that made the motors subject to flash-overs if the trolley voltage surged. The problem could have been solved by regearing the motors, but Wylie saw no need to do that. The plow ran fine when supplied by power from diesels, so diesels became the primary power source. The only exception was in the Bitterroots where cooperative substation operators sometimes reduced the voltage to 1,600 in order to make overhead power supply practical.

Soon after conversion, the 212 proved the value of electric drive. A major storm hit the Pacific Northwest on February 24, 1954. Over a period of nine days, six feet of snow fell in the Cascades and only the 212 was available to fight it. Due to its high horsepower, it was able to plow faster than a steam rotary and since it did not require fuel, water or maintenance, it could operate 16 hours a day every day. It performed a job which normally required three steam plows and when the storm passed, it was still running. Two of the three steam plows in the Bitterroots failed during the storm.

The Tacoma shops completed a conversion of the 900215 in 1955, while the Milwaukee shops completed the 900207 and 900214 in 1957. The plows converted in Milwaukee used diesel traction motors and operated only on 600 volt current.

After these plows were completed, all steam rotaries were stored. They never ran again and were scrapped in the early 1970's. Electric plows 212 and 214 were assigned to the Rocky Mountain Division while the 207 and 215 worked the Coast. Generally one plow on each division was held in reserve in case the other was damaged.

Steam rotary and crew after a losing battle with a snow slide filled with trees and rocks in 1922./Photo from Adam Gratz Collection.

X900212 soon after conversion to electric power. She is being tested in heavy wet snow at Hyak./Photo by Laurence Wylie.

A wedge plow sits in the snow at Three Forks, Montana, in March 1973. This type of plow was called a "Kiley Plow."/Photo by Noel T. Holley.

THE ELECTRICAL DISTRIBUTION SYSTEM

The Milwaukee's electrical distribution system included 654 route miles of trolley and feeder wires, 22 substations, a high voltage transmission line and approximately 40,000 wood poles. It was designed in 1914 to supply a nominal 3,000 volts DC to large locomotives in heavy service.

The type of catenary used on the main line was called Double Trolley. It was a simple catenary with a single stranded steel messenger wire and two side by side 4/0 grooved copper contact wires. Two trolley wires allowed the overhead to carry very heavy currents. Side by side installation meant that a pantograph could collect heavy currents without excessive arcing. Trains with several locomotives often drew currents ranging from 3,000 to 4,000 amps.

The standard height of the contact wires was 24 feet 2 inches above the rails. This allowed enough room for brakemen to give hand signals from the car tops without coming into contact with the trolley. Lower wire heights were found in places where the railroad passed through bridges, underpasses and tunnels. The original minimum height was 18 feet 9 inches. During the 1950's, this was increased to 19 feet in order to accommodate tri-level auto racks. Gaining this extra clearance required lowering the floors of several tunnels.

The standard lateral alignment for the trolley was anywhere within nine inches of a point above track center. Tests showed European style "staggering" to be unnecessary. The trolley did not wear a groove in the pantograph shoes because the normal rocking of the locomotives and the liberal alignment of the trolley spread the wear.

Most of the Milwaukee's trolley supports were tapered, 40 foot, red cedar poles. The originals were uncreosoted and by the mid-1920's, many had rotted badly below ground level. Between 1925 and 1937, they were bolted to creosoted fir stubs in order to extend their life.

The use of stubs was so successful that most of the pole line was not slated for replacement until the late 1970's. All replacement poles had creosoted butts in order to inhibit rot. Steel supports were used only in Butte, Renton, Seattle and Tacoma. Aesthetics was the primary reason for their use.

Standard pole spacing was 150 feet, but in areas of curving track and through yards, it was as little as 90 feet. This allowed the Milwaukee to align the trolley close to track center without excessive use of backbone wires and pull-offs. In order to limit the amount of guying needed on the pole line, the Milwaukee canted its poles back by an amount that was generally one-half inch per foot. This allowed the weight of the poles to counterbalance some of the weight of the trolley wires. Canted poles were difficult for linemen to keep in alignment, but there was a trade off. Vertical poles would have to be heavier, set farther from the track and steadied by more guy wires.

Due to rugged construction, the Milwaukee's trolley never suffered serious problems from storms or ice loading. It could carry a large amount of strain without breaking. For the most part, ice never built up very thick because the pantographs would knock it down. At Cedar Falls, however, ice sometimes interfered with operations.

> According to Laurence Wylie, "Once every few years, the sleet at Cedar Falls would ice up the wires to the point that pantographs couldn't make contact. We had to have men with clubs beat the ice off the wires in order to operate."

Maintenance on the trolley and the pole lines was a year-round task consisting primarily of inspection and making adjustments. The trolley expanded in the heat of summer and had to have slack drawn out. In the winter, it had to be loosened when it contracted in the cold.

Bipolar E1 silently speeds past Janney Substation as it ascends the eastbound grade to Pipestone Pass. Inside the brick building are three 1,500 KW M-G sets. Within the new metal addition is an ex-Cleveland Union Terminal 3,000 KW set. Date of photo is unknown./Photo courtesy Milwaukee, Wisconsin, Public Library.

Lateral alignment had to be maintained and poles had to be lined up or have their guys tightened. Lack of attention by inattentive or often understaffed trolley crews could result in pantograph dewirements and damage to the trolley.

Some wire alignment problems were not caused by weather or lack of attention by trolley crews. They were caused instead by shifting track. As Barry Kirk points out:

> "If trains go through a superelevated curve too slow, the inside rail often sinks. If trains go through any curve too fast, the outside rail sinks."

Because one inch of change in the elevation of a railhead will cause five inches of change in the location of track center 24 feet above the rails, poor track maintenance could cause problems. Another source of problems was maintenance-of-way crews who sometimes raised track or lined it up a few inches to the side during ballasting. That did not cause problems on unelectrified track, but in the electric zones, it left the trolley out of line and created work for the trolley crews.

Only three crews were needed to maintain the Milwaukee's trolley. Two were stationed on the 440 mile Rocky Mountain Division and one on the 214 mile Coast Division. Each crew consisted of a foreman, three linemen, and several laborers.

In the beginning, trolley crews worked from platforms built on top of boxcars. The boxcars were hauled by a steam engine so that work could be performed under dead wires. Steam trains proved to be slow and expensive, so the Milwaukee quickly replaced them with gas-electric cars called Troubleshooters. The Troubleshooters were equipped with hydraulically raised platforms for wire work. Inside, they carried wire and parts as well as bunks and a kitchen for the crews. In addition to the trolley crew, they carried an engineer, a brakeman and a conductor.

Barry Kirk, who worked his way up through trolley maintenance, said the following about the Troubleshooters:

> "They provided a good platform to work from, but they spent too much time in the shops to be dependably on call. In addition, they could not run around wrecks or quickly get in the clear when trains were coming. This made repair work slow.

The Troubleshooters were used until 1949 when Laurence Wylie replaced them with road-rail trucks. This move was part of Wylie's modernization program and it was a very significant move. Ten trucks were stationed at scattered locations along the line to replace the three gas-electrics. The trucks could reach any location by road or rail, they could drive off the track to get clear of trains and they could drive around wrecks. In addition, they did not re-

Troubleshooter X935019, one of the three original gas-electric trolley maintenance vehicles built by GE. Photo date and location unknown./Photo from General Electric Collection.

Troubleshooter X670 at Butte Yard, May 15, 1954. She saw only occasional use after the road-rail trucks were introduced in 1949./Photo by Wally Swanson.

X670 was built by Pullman in 1930. It was powered by a 400 hp Winton engine and GE electrical equipment. This car was 76 feet long and contained four bedrooms, a kitchen and a dining room. The photo location is Maple Valley, Washington, August 1956./Photo by Stan Styles.

This scene, just east of Snoqualmie Tunnel, shows typical Milwaukee Road trolley construction. Cedar poles with bracket arms supporting double trolley type catenary./Photo by Laurence Wylie.

In heavy snow areas, the Milwaukee used substations with gabled roofs. The snow here at East Portal is almost one story deep and the houses of company employees are already buried./Photo from Adam Gratz Collection.

Instructions to Hyak Substation operator dated August 1959: "Hold 3400 volts for all trains motoring when load conditions permit, except not necessary to hold more than 3000 for eastbound trains after arrival at Hyak. Hold 3200 volts for passenger train regeneration...Hold 2900 volts for 4000 ton freight regenerating."/Photo by Edwin Johanson.

quire a train crew. Wylie designed the first group of trolley trucks and was quite proud of them.

> "The easiest part of fixing up a wreck was putting the wires back into place," said Wylie. "Using trucks, the trolley crew could repair the wires and go back to bed before the wrecking crews even reached the wreck site."

Wylie's trucks had hydraulically raised ladders for the linemen to work from. Later trucks designed by Barry Kirk had hydraulic booms and work buckets. Some booms were powerful enough to set 90 foot poles.

Large brick substations housing transformers, oil switches, motor-generator sets and an operator's office fed current into the trolley wires. Indoor transformers were used because they were substantially cheaper than the outdoor type when the electrification was built. Motor-generators were used because mercury arc and solid state rectifiers had not been invented in 1914.

M-G sets convert AC to DC mechanically by using synchronous AC motors to turn DC generators. Substation operators could vary the generator voltage from zero to the maximum and they varied it to suit operating conditions.

The locomotives pulled more smoothly when operating on low voltage. They ran faster with high voltage. If a train was drawing so many amps that the generators were overloaded, the substation operator could protect his equipment by lowering the voltage.

This reduced the load significantly. A minimum of 1,800 volts was required in order to operate locomotives. Trains were generally supplied with current by more than one substation at a time and operators would vary their voltage to equalize loading or shift loads from one substation to another.

Whenever trains utilized regenerative braking, the locomotives generated a higher voltage than the substation M-G sets. The excess voltage drove the generators like motors. This forced rotation would cause the synchronous AC motors to generate alternating current. This current ran the kilowatt hour meters backward as it flowed into the AC power grid.

The Milwaukee's portion of the AC power grid was a 100 KV, 3-phase transmission line which paralleled the right of way. It interconnected the substations and served as a tie-in with the electric utility companies which supplied the power. On the Rocky Mountain Division, the supplier was Montana Power Company. On the Coast Division, it was Washington Water Power on the east end and Puget Sound Power & Light in the west. For many years the Milwaukee's "high-line" was the only inter-tie between the two Washington based utilities.

When the Milwaukee first negotiated contracts with its power suppliers, estimates were that electrical system loading would average 60% of capacity. As a result, the Milwaukee agreed to a rate structure where the minimum payment reflected consuming that much power. The payment was to rise if peak power consumption exceeded that level.

Because the power companies had limited generating capacities, they feared that huge current demands from Milwaukee trains could have a disruptive impact on their other customers. In order to limit this, and to protect the railroad from high power bills, the Milwaukee installed a Power Indicating and Limiting system. The P.I.L. was in place by 1920, but only operated a few years. The Milwaukee's traffic never grew enough to use the amount of current reflected in the power bills and thus it was never disruptive. With agreement from the power companies, the Milwaukee dismantled the P.I.L. and renegotiated its power contracts to reflect realistic traffic levels. Thereafter, Montana Power and Puget Sound Power & Light charged a flat rate per kilowatt hour. Washington Water Power also charged a flat rate, but that rate could rise to reflect the amount of current used during peak load time. That time was 5:00 p.m. to 6:00 p.m. during the period from October 16th to February 15th. The Milwaukee's rates averaged $.00544 per kilowatt hour.

Although the low number of trains kept total power consumption low, individual trains could draw large amounts of current. The Milwaukee's substations were designed to supply power to 1915 era freight trains while holding one M-G set idle as a spare. As train size grew, they were able to meet the demands by running all of the M-G sets and to some extent by relying on overload capacity. The M-G sets could provide 150% of rated power for up to two hours.

By the end of World War II, the limits of the power supply system had become a real concern. Trains weighed as much as 5,000 tons and diesels could best the performance of electrics on tonnage trains. Not only were the electric locomotives old, but the electrical distribution system was having difficulty providing the power needed to keep up. If newer, more powerful, locomotives were purchased, they would place an even greater burden on the overloaded supply system.

Rainier Beeuwkes, who headed the electrification until 1948 felt that the only way to alleviate the problems would be to make a massive investment in new electrical distribution equipment as well as new locomotives. He was not, however, totally in favor of that move. Some experts in the area of electrification questioned whether or not the required amount of current was practical to supply at 3,000 volts. Additionally, most experts including Beeuwkes rejected the idea of a voltage increase as being too risky. In the face of these problems, Beeuwkes concluded that the Milwaukee's electrification was simply outmoded.

An Electrification Department official inspects a new tower car built by the Tacoma shops. The crowded track car has just towed it to this spot near the Maple Valley depot./Photo by Laurence Wylie.

T1 is shown at Doris in 1954 where crews are repairing damage from the E41 wreck. This vehicle was designed by a Coast Division trolley foreman and built in the Tacoma shops./Photo by Laurence Wylie.

This bright red Dodge road-railer with hydraulically raised ladders was built under the direction of Laurence Wylie./ Photo by Laurence Wylie.

When Laurence Wylie succeeded Beeukes in 1948, he immediately went to work implementing a number of low cost changes which allowed the existing system to meet the needs and operate more economically. Among the first was a voltage increase. Substation maximums were raised from 3,200 to 3,600 volts, raising typical line voltages to the 2,800 to 3,300 zone.

The amount of power supplied by the system was a function of both amps and volts. It was preferable to increase power by increasing amperage capacity with more and bigger substations, but economics eliminated this as an option. As a result, Wylie looked at voltage. Even a 10% increase would mean the delivery of 21% more power. This would work against the line losses that sometimes brought the voltage down to as little as 2,000 volts between substations, and it would speed up the trains.

Wylie's low cost proposal to increase voltage by shimming the air gap between the generator rotors and stators was vetoed by the railroad as unsafe. Undaunted, Wylie proceeded in 1950 by quietly modifying the M-G sets on the Coast Division. After proving out his theories in operation, Wylie used this success to argue for a voltage increase on the Rocky Mountain Division. Since it worked, permission was granted to proceed with the modifications.

Retired Coast Division substation operator Bill Bibby recalls: ''We were on pins and needles when they reduced the gap because we expected a lot of trouble. If any of the field coils raised up above proper height, they would have shorted out the generators. Things worked pretty well though. We kept a close watch on the coils and most of them stayed tightly in place. After reducing the gap, you could get the voltage up to 3,500 or 3,600. We didn't generally go higher than 3,400 because above that we were getting into dangerous territory.''

Substation automation was another improvement initiated under Wylie. Each substation originally required three operators for 24-hour coverage. Remote control allowed operators at one substation to control the operation of several distant substations. The result was a great reduction in labor costs. The automation program was carried out by existing staff on a low budget. It progressed slowly, continued over a long period of time and was never totally completed. It was begun by Earl Barnes and George Frazier who were two of Wylie's assistants. Barnes developed several low cost methods for achieving remote control operation. Tacoma substation was used for testing ideas, and it was made totally automatic in 1953. The first prototype remote control system was installed on the Rocky Mountain Division. It allowed Tarkio operators to run Primrose. The last system installed allowed Cedar Falls to run Renton. Three was the largest number of substations which could be controlled from one spot. According to Laurence Wylie, groupings larger than three would have required the purchase of expensive control equipment which was not otherwise needed.

Above: The trolley crew adjusts the wires from atop Troubleshooter X671. From left to right, they are Lineman McClure, Lineman Tom Fairhurst and Trolley Foreman Fred Kirk./Photo from Fred Kirk Colleciton.

Left: Barry Kirk was a pipe smoking lineman in 1938 when he posed with Duke Seguin, the cook on Troubleshooter X501./Photo from Fred Kirk Collection.

Transformers and oil switches at Hyak substation August 25, 1962./Photo by Gordon Rogers.

The substation operator and his assistant stand at the Cedar Falls control panel. The date is approximately 1920./Photo by Laurence Wylie.

One of the big 2000
motor-generator sets
Coast Division sub
tion. A synchronous
motor sits between
1500 volt DC generat
The two generators
wired in series to prov
3000 volts./Photo
Laurence Wylie.

"Fireman" for the Milwaukee electrification—it's a term well-suited to Bob William's role as substation operator at the key mountain top generating facility at East Portal. Here's a portrait of Bob in the transformer room at East Portal in May, 1974./Photo by Ted Benson.

"GE and Westinghouse proposed replacing all of the substation equipment with modern designs and automating their control," said Wylie. "We looked at it, but the changes would not have saved us enough money to be worthwhile."

In a continuing effort to keep costs low, Wylie negotiated lower rates on the power contracts. In addition, he kept an eye on the used equipment market in hopes of finding cheap but useful electrical gear. This paid off in 1955 when the Cleveland Union Terminal de-electrified. Their five huge 3,000 volt motor-generators went on the block, and Wylie bought two of them. One was installed at Janney; it increased that substation's capacity from 4,500 KW to 7,500 KW. The other M-G set went to Cle Elum, and Cle Elum's original set was then moved to Doris. These moves raised the capacity of Cle Elum from 2,000 KW to 3,000 KW and Doris from 4,000 KW to 6,000 KW. Doris and Janney both fed current to trains on grueling mountain grades. The additional capacity was sorely needed in both locations. Additional feeder wire helped to distribute power in areas that were hard-pressed.

MILWAUKEE ROAD SUBSTATIONS

Substation Number	Substation Name	Control System	Number of M-G Sets	Kilowatt Rating
1.	Two Dot	• Manual & Remote	2	4,000
2.	Loweth	• Manual & Remote	2	4,000
3.	Francis	• Manual & Remote	2	4,000
4.	Eustis	Manual & Remote	2	4,000
5.	Piedmont	Manual	3	4,500
6.	Janney	Manual	4*	7,500
7.	Morel	• Manual & Remote	2	4,000
8.	Gold Creek	• Manual & Remote	2	4,000
9.	Ravenna	• Manual & Remote	2	4,000
10.	Primrose	• Manual & Remote	2	4,000
11.	Tarkio	• Manual & Remote	2	4,000
12.	Drexel	• Manual & Remote	2	4,000
13.	East Portal	Manual	3	6,000
14.	Avery	Manual	3	4,500
15.	St. Joe	---------NEVER BUILT		
16.	Plummer Jct.	---------NEVER BUILT		
17.	Rosalia	---------NEVER BUILT		
18.	Castleton	---------NEVER BUILT		
19.	Ralston	---------NEVER BUILT		
20.	Roxboro	---------NEVER BUILT		
21.	Tauton	Manual	2	4,000
22.	Doris	Manual	3**	6,000
23.	Kittitas	Manual	2	4,000
24.	Cle Elum	•Manual & Remote	1*	3,000
25.	Hyak	Manual	2	4,000
26.	Cedar Falls	Manual	2	4,000
27.	Renton	Manual & Remote	2***	4,000
28.	Tacoma Jct.	Automatic	1***	2,000

• Substations that could be operated either manually or by remote control.

Manually operated substations that could control others remotely.

* One ex-Cleveland Union Terminal M-G set installed in 1955.

** Doris built with two M-G sets, received a third from Cle Elum in 1955.

*** Tacoma was built with two M-G sets, Renton was built with one. During the 1920's, one set was moved from Tacoma to Renton.

Substation capacity: 100% of rating continuously, 150% of rating for two hours. 1 kilowatt 1.36 horsepower 1 HP = .746 KW. Because of resistance losses in the trolley, kilowatts delivered to the locomotive were somewhat lower. The amount depended on the distance from the locomotive to the substation.

Data provided by Electrification Departme

Williams checks the oil cups in his MG sets, making sure they'll have sufficient lubrication to carry another heavy load over the mountain. May, 1974./Photo by Ted Benson.

After voltage and capacity were increased, operations went smoothly, but cooperation was required between engineers and substation operators. Engineers had to make an effort not to draw too many amps, while substation operators provided the most possible power by feeding trains from more than one substation. As a rule of thumb, the maximum amperage which could be supplied to one train was 4,000. One Little Joe being worked hard could draw 3,000 by itself. Trains on the Rocky Mountain Division generally ran with two Little Joes and in the mountains a four-unit GE Motor was added to the train as a helper. On the Coast Division it was all GE Motors, but they too could draw a large amount of current if worked hard. Both engineers and substation operators watched their ammeters closely to prevent overloading.

East Portal substation routinely supplied 2,600 volts to trains with Double Joes and helpers in order to avoid the overloading that came at higher voltages. At Cedar Falls, Frank Zellerhoff often gave low voltage to GE Motors.

> "Engineers going by would give us the thumbs up signal for more voltage," says Frank, "but they didn't realize that they already had the substations loaded up 3,200 amps. If we gave them any more voltage, they would trip us off the line. Some engineers never understood how the electrical system worked and since the substations didn't have radios in them, we couldn't tell those guys they were overloading us."

Through cooperation, the Milwaukee was able to run trains weighing in excess of 6,000 tons. Occasionally, however, there were people who did not make an effort to cooperate and retired Coast Division engineer Pat Chester remembers some of them.

> "Some engineers would fight with the substation operators. You couldn't win when you did it, but some guys were ornery and they would do it anyway."

June 13, 1974 finds train #264 eastbound at the 4000 kilowatt Primrose substation. E72, 4005, 2058, 16, 4008 in the lashup./Photo by Dick Dorn.

"You might go by a substation and see the operator in there reading a book or out watering his garden. Then you would get out between substations and your motor would only be going 30 mph instead of 40 mph because the guy wasn't paying attention and let the voltage drop down low. Well that would make you mad so you would just kick the throttle up a couple of notches and overload him. That would flash the commutators on his M-G sets and trip out his breakers. Those guys hated that because those commutators were so big it would take hours to clean the carbon off of them. Next time you went by his substation, he would be shaking his fist at you. He might even give you low voltage on purpose or make it hard for you to get into regeneration by 'forgetting' to lower his voltage when you were going down grade."

"We used to trip out the substations too," says retired Rocky Mountain Division engineer Adam Gratz, "then the Electrification Department put time delays in the substations. Whenever the substation was tripped out, the delay kept the operator from coming back on line until a certain amount of time had passed. We stopped tripping them out after that. We had to sit around so long we were hurting ourselves worse that we were hurting them."

Because the basic problem underlying trip outs and low voltage was lack of capacity, the Electrification Department made plans to increase power supply. These plans involved building more or larger substations, adopting solid state power conversion which would operate under centralized control, and raising the voltage as high as 3,750 volts.

The question of upgrading the power supply system was, because of its expense, part of a much larger debate. The entire electrification was in need of costly upgrading and a decision needed to be made on whether to keep it and invest in it or to scrap it. The trolley wire had many years of life left in it, but the poles would have to be replaced. New locomotives were immediately needed to replace the old GE Freight Motors and the Little Joes would need replacement in a few years. The estimated cost of these improvements was $39 million.

After much study and debate, the Milwaukee's management concluded that the costs of re-electrification exceeded the benefits. Additional diesels were purchased instead and in 1973 the removal of copper wire began. The copper recovery work took 3½ years, and when the costs of recovery were deducted from the sale of scrap, the railroad netted $7 million. Most of the substation buildings were sold to used brick dealers and torn down.

On November 12, 1978, headlight extinguished for a meet with Extra 559 east, Extra 167 west rolls 'em toward Beverly at dusk under skeletal remains of the electrical system. It's dusk for the Milwaukee Road, as well. A full moon is supposed to be ominous, predicting a foreboding future. In this case, it is./Photo by Blair Kooistra.

IDAHO DIVISION

The Unelectrified Gap

By 1917, the 440 mile Rocky Mountain Division electrification was in full operation. This segment, between Harlowton, Montana, and Avery, Idaho, comprised approximately one-half of the Milwaukee's proposed electrification. The other half was the 436 miles of track connecting Avery and the port city of Tacoma, Washington. This trackage included the 212 mile Idaho Division mainline and 224 miles of mainline on the Coast Division.

As Chief Electrical Engineer Rainier Beeuwkes mulled over which division to electrify first, several factors steered him away from the Idaho Division. Traffic was split over two routes. Passenger trains traveled a 225 mile route which passed through Spokane, while freight took the 212 mile mainline through Malden. Neither line had sufficient traffic to utilize an electric power distribution system very efficiently. Additionally, 82 miles of the passenger route was on Union Pacific track. The Milwaukee was already paying a wheelage rate for trains using U.P. track and if it chose to electrify that route it would also have to pay an enormous fee for the right to put up trolley wires. Thus trolley was proposed only for the freight line. Because this contained no heavy grades, there was no expectation that a dramatic improvement in service would result from substituting GE Motors for steam engines. Beeuwkes felt that if he electrified the Idaho Division and it was not an economic success, he might never get permission to electrify the Coast Division. On the other hand, the Coast Division was sure to be an economic success, which would add leverage to the argument for completing the total electrification project. In addition, if both the Rocky Mountain Division and the Coast Division were electrically operated, it would seem only logical to link them and improve operational efficiency.

In 1918 crews began electrifying the Coast Division and by 1920 it was completed. The project was a success, but due to post-World War I inflation, the cost was significantly higher than the estimates. That cost, plus the debt-weakened financial position of the railroad as a whole led to postponement of the Idaho Division electrification work. In 1924, the Milwaukee declared bankruptcy and by the time it was reorganized, the Great Depression gripped the land. Due to continuing financial problems, World War II and then the arrival of diesel locomotives, the Idaho Division was never electrified. No wires were strung and the substations assigned numbers 15 through 20 were never built. The substation plots at St. Joe, Plummer Jct., Rosalia, Castleton, Ralston, and Roxboro sat empty and the division became known as "The Gap."

During the 1950's and 1960's, the Milwaukee studied electrification of The Gap several times. The conclusion was always the same. At current traffic levels electric operation would not save enough money to justify itself relative to diesel operation.

Electric operations terminated at each end of the Idaho Division, Avery, Idaho on the east and Othello, Washington on the west. Between these points lay a fiefdom of steam and when steam was no longer king, it was ruled in turn by diesels.

The Gap was known for its high train speeds. Most curves and grades were gentle and the track was well-maintained until 1961 when passenger sevice was ended. On the line from Avery, through Malden, to Othello, speed limits varied from 50 to 70 mph for passenger trains and 40 to 50 mph for freight. The only exceptions were in the winding curves near Malden and Avery. Operating speeds were often near the limits and Pat Chester, who worked on the Idaho Division, recalls that the limits were often held in the same regard as those on the highways.

> "If a guy was late, those went by the wayside. He was way up over the limit, sometimes 80 mph."

On most of the line through Spokane, the speed limits were 70 mph for passenger and 50 mph for freight. This track was jointly operated with the Union Pacific and the maintenance standards were high. The portion of the line from Manito through Spokane, to Marengo Jct. was owned by the U.P. Manito to Spokane was part of the Tekoa Branch. The 65 miles of line between Spokane and Marengo Jct. was part of the Spokane to Hinkle mainline. Union Pacific mainline construction was exceptional by Milwaukee's standards and Milwaukee crews sometimes called it "The Freeway."

> "There was no comparison between U.P. track and Milwaukee track," says retired engineer Paul Edmunds. "You could make up time on the U.P. where you couldn't on the Milwaukee. If you left Marengo 10-15 minutes late you knew you would be on time when you got to Spokane. The U.P. was tough on the rules though. They were always after some of our men for speeding."

The story of The Gap is the story of non-electric motive power and that is the story which follows.

The 251 and train #18, the Columbian, approaching St. Maries in 1953. The stack extension has been raised to keep the smoke up./Photo by Jim Fredrickson.

OIL IN THE FIREBOX

Steam locomotives on the Milwaukee Road were predominantly coal burning types. Very few of its steamers burned oil except in the Pacific Northwest. When the Pacific Extension was completed in 1905, coal was used all the way to Tacoma. Within the next fifteen years, however, a conversion took place. Oil came into use west of Three Forks. Steam on the Idaho Division, the Coast Division and most of the Rocky Mountain Division became exclusively oil fired.

> "For a long time we could use coal burners between Avery and Malden, but not east of Avery," says retired engineer Bill Plybon. "They didn't want to run the risk of starting fires in the big timber east of Avery. On the Rocky Mountain Division they converted to oil as soon as they got bigger engines. The first ones we had were just small engines, hand fired. I think it was around 1920 when they converted the Idaho Division to oil. They did it so they could use our engines east of Avery if they wanted to. The engines steamed much better on oil than on coal. Oil was much more dependable. Even with mechanical stokers you still had holes in your firebed that you had to fill with a shovel."

Using oil as fuel required some changes in the locomotive firebox and it also meant changes in the fireman's job. The back-breaking work of shoveling coal for 16 hours was gone, but the oil burning equipment required a lot of attention to function properly.

The fireman's job was to watch over the fuel, water, and lubrication of the engine and to give the engineer all the steam pressure needed to roll the train. In addition to controlling the oil flow with the firing valve, the fireman operated the damper, which admitted air to the firebox, and the steam heater, which kept the oil hot in the tender.

The Milwaukee, along with many other American railroads, used heavy residual fuel oils as locomotive fuel. Residual fuel is the substance which remains after gasoline, kerosene, and

4-6-0 locomotives were withdrawn from the Idaho Division as soon as heavier power was available. However, some remained on the Rocky Mountain Division branch from Three Forks to Bozeman. On July 4, 1951, G Engine 1029 left Three Forks with a four car train./Photo by Phillip R. Hastings.

An engine change in Othello, November 1, 1950. "S Engine" #267 prepares to haul the Olympian Hiawatha on its nighttime journey across The Gap./Photo by Philip R. Hastings.

Mallet #50 at the Clarkia depot after bringing in a train from Bovill. October 14, 1950./Photo by Philip R. Hastings.

At the east end of The Gap, 4-8-4 #250, 4-6-4 #131 and 2-6-6-2 #65 await their trains. Oil smoke and haze hang lazily above the Avery engine terminal on 9-2-50./Photo by Philip R. Hastings.

light oils such as diesel have been removed from crude oil. Bunker C is the cheapest and heaviest of the residual fuels. It is a tar-like substance which is not truly a liquid unless heated. It does not burn well until heated to its flash point which is between 170 and 180 degrees. The grade above Bunker C is a liquid, but to burn well it must be heated to a point near 120 degrees. Bunker C, when burned produces 12% more heat per gallon than diesel oil, and it is substantially cheaper than diesel oil, enough so that its price is competitive with coal when used as a boiler fuel. It is, however, hard to work with because of its tarlike consistency and the type of equipment used for burning it.

The coal grates were removed from the locomotive firebox and were replaced with a floor of firebricks. The brickwork extended part of the way up the sides of the firebox. In the front of the firebox was an atomizer. Within it, pre-heated oil was put under pressure by mixing it with steam. The pressure propelled a fine spray of oil through small holes in the atomizer. The oil sprayed rearward in the firebox toward a flashblock made of firebrick. The flashblock was located below the firebox door, and once it was hot it served to ignite the oil.

To start a fire, the fireman tossed kerosene soaked cotton waste into the firebox and lit it. Then a small spray of oil was fed through the flames and burned. An oil fire is hot enough to melt the grates out of a coal burning engine, and it heated the firebricks quickly. When the bricks got hot enough to ignite oil, the fuel flow could be increased. From that point on, the oil would ignite from contact with the firebricks in the flashblock. Firebricks retain heat and the roaring fire kept them hot.

The operational differences between oil burners and coal burners is illustrated through the memories of Ralph Danley. Danley spent most of his career firing coal burning steam engines on the Great Falls branch. During the coal strike of 1949, the Milwaukee sent in oil burners and Danley recalls some of the things firemen needed to learn in order to make the transition.

> "There was no bed of hot coals and ash to warm up cold air and keep it from coming into the firebox and hitting the flues. You controlled the temperature in the firebox by opening and closing the damper. If cold air hit the flues it would shrink them and break them loose from the flue sheet. Then you would have a steam leak. Some guys had a lot of problems with that. Once I fired an engine that had a water leak as big as your arm coming out around a leaky flue.
>
> If the engineer was going to slow down you would close the damper first and then shut off the oil. If you shut off the oil first, you would have no fire to keep the cold air out. If he was speeding up, you would open up the oil a notch for every notch he opened up the throttle. You would wait until you saw a little smoke at the stack and then open the damper a notch. You burned a whitish orange flame and it was so bright you needed sun glasses to look at it. You could get instant heat with oil, much faster than coal."

Pat Chester was an Idaho Division fireman. He often fired on the mainline where big engines and 70 mph speeds were a routine part of a demanding job. With the following vignettes he tells how it was.

> "You had to learn a lot to fire steam engines. If you didn't know how to keep the engine hot the engineers wouldn't run with you and you would be out of work. You had to know each engine, know what each engineer was going to do, and anticipate everything.

DRAWING SHOWING TYPICAL ARRANGEMENT OF MILWAUKEE ROAD OIL BURNING EQUIPMENT

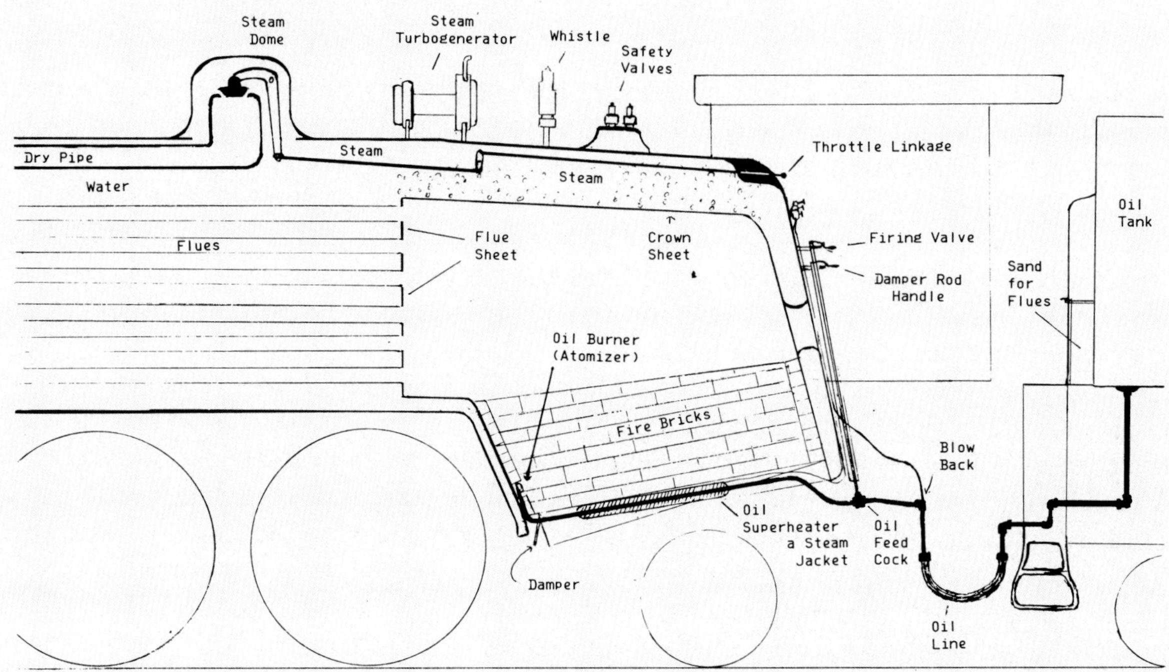

OIL BURNER SPRAYS HOT OIL ON THE FIRE BRICKS BELOW THE FIRE DOOR. THESE BRICKS ARE REFERRED TO AS THE FLASH BLOCK.

NOT TO SCALE

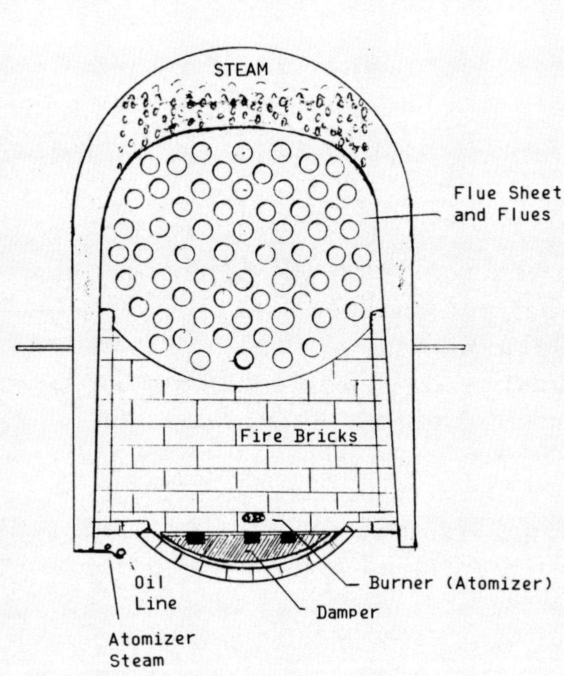

Firebox sides and floor are lined with fire bricks

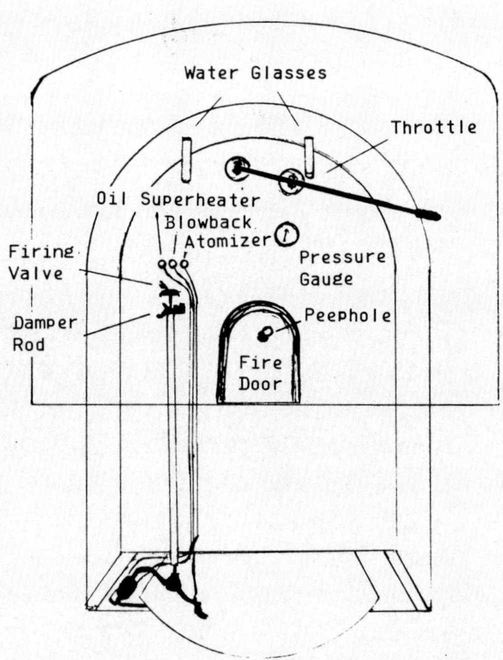

Damper rod is a pipe with the firing valve rod down its center. Firing valve rotates to control feed cock. Damper rod is raised and lowered to control damper linkage. Damper rod handle rests on a quadrant having steps at different heights. Damper swings outward.

The oil in the tender had to be kept hot so it would burn right when it got to the firebox. Some tenders had short heat pipes in them. They had gotten plugged up at one time or another and the shop crews just cut them off instead of fixing them. If the tender on your engine had a short heat pipe, you might have to turn the steam way up to get heat all the way to the bottom of the oil bunker. That same setting on a tender where the pipe came all the way down would overheat the oil. The only way you knew which tender was which was to remember them.

Another thing you had to think about was when you were going to take on water. The water in the tender sat around the oil bunker like a horseshoe. Some of the tenders held 18-20,000 gallons of water and if you put 18,000 gallons of cold water around your oil bunker it would cool the oil down. If you didn't have the oil extra hot when you took water, it might be a while before you got the oil hot enough to burn right again.

You had to be ahead of the engineer with the firing valve. You had to have it open and make smoke before he opened up the throttle. If you waited until he opened up the throttle before you opened up the oil, you were already behind him. You might get the steam pressure back up after a while and you might not.

Each engine had a damper in the firebox. The N class mallets had two, one in the front and another one under the cab. If you didn't work them right or your oil was cold, you could get a big pile of carbon building up in front of the firebox door. Carbon is unburned fuel and it would cover up the firebricks. It would keep the bricks from getting hot and then your engine wouldn't steam well because the fuel wouldn't burn well. Sometimes the carbon would build up in a pile 2 or 3 feet high. You would have to chip it out with bars and chisels to get rid of it.

You had to learn the line and you had to learn what each engineer was like. Some engineers were very demanding and some weren't. One engineer I fired for was always cautious and careful. He didn't want to break anything or take any chances. One time the conductor came up and told him, "I only seen three sparks come out of your smokestack the whole trip and you were going so slow that two of them fell back in." This guy's brother was just the opposite. He was a fast runner and he would work the engine hard. He figured that if something broke, then someone else would have to fix it. If any sparks came out of his smokestack they were liable to come down in the next county.

In order to keep steam up you had to learn how each engineer operated. Usually they wouldn't tell you anything, you would just have to learn by experience. I remember one old engineer who gave me a hint though. He said, "Just remember kid, when I put my glove on it don't mean I'm about to pick my nose." Anytime I saw him put on his glove, I opened up the firing valve because I knew he was going to yank the throttle back.

Some of those guys were all business when they came up into the cab. One night I was firing for an old engineer who never talked to you, he just ran the engine. He expected you to know what he was going to do and to keep the steam up. Well I fell asleep. Next thing I knew he was holding me up by the scruff of the neck and telling me, "Kid, if I so much as see you blink an eyelash again, I'm going to throw you off of this engine." I was so scared I didn't sleep again for three days.

Those old engineers knew a lot and you had to respect them. A lot of them worked their way up through the machinist's ranks and they knew everything there was to know about a steam locomotive.

One time I was firing for one of them and we were out about sixty miles from nowhere when the throttle linkage broke somewhere inside the boiler. Do you know how we got home? He ran the locomotive with the Johnson Bar. When it was hitched up in the center, no steamed flowed. When he dropped it down a little, it let enough steam into the cylinders to get us rolling. He controlled the speed that way and got us back in. If he sat and waited for someone from the roundhouse to come fix it, who knows how long we might have been out there."

A slobber-stack at Tacoma Yard 4-10-40. The 92 was a class N-2 compound mallet 2-6-6-2./Photo by Albert Farrow.

A southbound freight on the Metaline Falls Branch crosses over the Great Northern mainline near Newport, Washington in September, 1950./Photo by Philip R. Hastings.

At Avery yard, Freight Motor E55 and F7 diesel #88 make up their trains and prepare to leave town on September 2, 1950./ Photo by Philip R. Hastings.

THE SLOBBER-STACKS AND THE STRAIGHT-EIGHTS
CLASS N, 2-6-6-2's

Slobber-stacks is what crews called the N-1 and N-2 mallet compounds. Many of these engines were not equipped with super-heaters and by the time their saturated steam had passed through two sets of cylinders its temperature was low. It went up the stack in a whooshing exhaust that was cool enough to condense and fall back on the locomotive like dew.

"The Slobber-stacks were a pain in the neck to run," says retired engineer King Clover. "They leaked steam from everywhere and in cold weather you couldn't see in front of you, behind you or beside you. They were slower than the dickens and if you got up to 30 mph, you were moving right along. If you opened the 'dutchman' and put live steam into all of the cylinders they would lug pretty good, but you couldn't do that for very long. They would run out of steam."

Straight-eights was the name for the N-3 class, a well liked group of simple articulateds which the Milwaukee created by rebuilding N-1's.

"They were good engines," says Clover.

Bill Plybon, an Idaho Division engineer from 1919 to 1965 recalls how they got their name.

"We called them Straight-Eights because the exhaust from all four cylinders went straight up the stack and that gave you eight separate exhausts for every revolution of the drivers."

There were originally 41 locomotives designated Class N. They were built to haul freight through the mountains and they were the only articulated steam power the Milwaukee ever

Mallet #64 at Avery, Idaho before its Coffin feedwater heater was removed. Date unknown./Photo by Gayle Christen.

Most of the N-3 mallets had Coffin feedwater heaters. #64, shown here, had none. Tacoma Yard, April 11, 1954./Photo by Albert Farrow.

Straight Eight #63 waits to come off the Tacoma Eastern near the old Tacoma depot. May 10, 1944./Photo by Albert Farrow.

owned. Prior to electrification of the Milwaukee's mountainous disticts they were assigned to the Coast and Rocky Mountain Divisions.

The 2-6-6-2's came to the Milwaukee as two nearly identical groups of mallet compounds. Class N-1 was built from December of 1910 to January of 1911. It included 25 locomotives numbered 9500-9524. Class N-2 was built from May of 1912 to October of 1912. It was 16 locomotives numbered 9100-9104 and 9600-9610. All were built by American Locomotive Company, in Schenectady, New York. The N-1 and N-2 class compounds were 88 feet 7 inches long including tenders, and they rode on 57 inch drivers. The low pressure cylinders were 37 inches in diameter and the high pressure cylinders 23½ inches in diameter. Boiler pressure was 200 pounds per square inch and the tractive effort was 70,396 lbs. Only the N-2 class had superheaters. The N-1 group weighed in at 277 tons while the N-2's were four tons heavier at 281 tons. They were slow and ponderous machines, but by 1912 standards they were state-of-the-art steam engines for heavy mountain freight service.

Prior to electrification, the Milwaukee used its Class N mallets in helper service. When the wires went up, the mainline steam engines on the Rocky Mountain and Coast Divisions were displaced. Most were well suited to service on other parts of the system, but not the Class N mallets. Most of these engines became surplus. They were too slow and inefficient for use on the prairies east of Harlowton so the Milwaukee tried other assignments for them. Some were assigned to the Idaho Division mainline, some to branchlines on the Idaho and Coast Divi-

An N-1 compound mallet helps an F-4 class road engine and the eastbound Olympian in this pre-electrification scene at Cedar Falls. The un-superheated N-1 class had steam pipes directly from the steam dome to the high pressure cylinders. Date unknown./Photo by J.M. Holzman from the Allen Miller collection.

*The eastbound Olympian and Mallet 9301 prepare to leave Spokane's Union Station in April of 1932. This locomotive later became #51./*Photo by R.V. Nixon.

sions and some to mine runs at various points on the system.

The Class N compounds were not highly successful on any of these assignments. On branchlines and the relatively flat Idaho Division they could not make speed or handle heavy trains. Additionally they used excessive amounts of fuel and water when operated in general freight service.

Because of reliance on Class N Steam power the Idaho Division became a bottleneck for traffic. GE Motors could haul 5,000 ton trains across the mountains to Avery and Othello, but Class N mallets were incapable of rolling trains that size across The Gap.

One solution to the problem would have been the purchase of new locomotives to replace them. Steam engine design had taken great leaps forward and several faster, more powerful designs entered service on other railroads. In 1925, 2-8-4's, 2-10-4's and 4-10-2's were introduced while in 1927, the Northern Pacific built its first 4-8-4's.

The Milwaukee was in bankruptcy during the mid 1920's and it had difficulty financing the construction of new locomotives. The Mechanical Department did, however, take note of what resulted when other railroads rebuilt their old mallets as simple articulateds. There was generally a large increase in power and speed.

By 1928, corporate reorganization had eased the financial crisis. The Mechanical Department was then able to begin a motive power modernization program. As part of that program locomotive 9504, a Class N-1 was experimentally rebuilt. In January of 1929, it emerged from the shops as the first N-3. All cylinders were reduced in size to 21½ inch diamter and fed with high pressure steam. The evaporative capacity of the boiler was increased by installing a superheater, a Coffin feedwater heater, and four thermic syphons.

After several months of testing, the rebuilt locomotive was declared to be a success. It could pull 10% to 15% more tonnage than the old compounds and it could roll its trains almost twice as fast. The old N-1 and N-2 compounds ascended grades at 7-8 mph and had trouble exceeding 30 mph under any conditions. The N-3 could ascend grades at 15 mph and make 60 mph on a slight down grade. In addition, axle loading was still light enough to allow use on branchline trackage.

Beginning in July of 1929, the Tacoma shops converted 16 more N-1's to simple operation. New tenders were constructed for the fleet of 17 N-3 locomotives at the Milwaukee, Wisconsin, shops. These tenders were only ten feet longer than the originals, but they had almost double the oil and water capacity. They held 18,000 gallons of water, 7,350 gallons of oil, and rode on six-wheel trucks. Class N-3 locomotives complete with tenders weighed 370 tons and had 82,720 lb of tractive effort. The new class was numbered 9300 through 9316.

Because the Milwaukee still had more Class N locomotives than it needed, the 8 remaining N-1's and 13 of the N-2's were scapped between 1927 and 1935. Three N-2 compounds were retained on the Coast Division. These were 9601, 9602 and 9606.

The 20 Class N locomotives were split between the Idaho and Coast Divisions. Ten N-3's were based in Othello, while 7 N-3's and the 3 N-2's were based in Tacoma.

When steam locomotives were renumbered in 1938, the N-3 straight-eights became 50 through 66, while the N-2 Slobber-stacks were 90, 91, and 92.

On the Idaho Division, the N-3 locomotives were the heaviest power in service until the late 1930's, and they were the most numerous heavy power until the arrival of dieselization. Their work included both passenger service and freight, mainlines and branchlines. When passenger trains got too heavy for the light 4-6-2's the Milwaukee operated, N-3's were assigned to the job. They could pull a train of 19 cars where a 4-6-2 could only pull 8 to 10. Several of the N-3's were equipped with steam lines for train heating in order to provide for their use in passenger service. The most common use for N-3's was the through freight service between Avery and Othello and also working heavy trains on branchlines. The branchlines which most often saw them were Metaline Falls and Elk River.

On the Coast Division, the Class N locomotives were used almost exclusively on branchlines. They ran to Morton, Enumclaw and Everett. The N-3's could be found on any of those lines while the N-2's were generally found on the low speed branches to Enumclaw and Everett. During World War II, N-3's occasionally pulled passenger trains on the mainline. High wartime traffic sometimes produced more passenger trains than there were Bipolars. The Coast Division had no passenger steam engines assigned to it and the N-3's were first choice from among the ranks of freight steam engines. High wartime traffic created power shortages on several western railroads and among them the Western Pacific was particularly hard hit. To alleviate this shortage, the government allocated mallet #58 to the Western Pacific for the duration of the war.

After the war, the Class N locomotives continued to serve the Milwaukee until they were replaced by diesels. The N-2 Slobber-stacks were the first to go, being retired in 1949.

The Straight-Eights followed one by one until 1954 when number 64, the last Class N, was retired.

Straight Eight #50 climbs a grade on the Tacoma Eastern, a branch southeast of Tacoma, September 18, 1947./Photo by Albert Farrow.

THE F ENGINES
CLASSES F-4, F-4ms AND F-5, 4-6-2's

The F engines were the Idaho Division's general purpose steam locomotives. They worked in passenger service and freight, ran on branchlines and mainlines, and acquired a reputation as excellent light, fast, motive power. These 4-6-2's were built from 1910 through 1912 to haul passenger trains. Some were built by ALCO and others by the Milwaukee, Wisconsin shops.

System-wide, the Milwaukee once owned a total of 162 class F 4-6-2's. The first was built by ALCO in 1889 and designated class F-1. The second was built by the Milwaukee shops in 1905 and desgnated class F-2. Neither of these early locomotives saw use in the West. As the Pacific Extension was completed 160 new class F's were built. Class F-3 was designed for speed. These engines had 79 inch drivers and were assigned to the relatively flat Midwestern prairies. Classes F-4 and F-5 were built with 69 inch drivers and the power to operate on the mountainous line from Harlowton to Tacoma. After the electrification was completed most of the F-4 and F-5 engines were assigned to the Midwest. Just ten remained on The Gap. Seven of the ten belonged to class F-5b; they had a tractive effort of 43,116 lbs. Two were F-4b's with a tractive effort of 39,736 lbs. One was an F-4ms, a locomotive which had its driver tires built up to 73 inches. Although the F-4ms was potentially faster, its tractive effort was correspondingly lower at 37,558 lbs.

*F Engine #835, a light 4-6-2 sits near the Avery roundhouse. Date unknown. Engines of this type sometimes pulled the Butte Local from Butte to Avery. The train left Butte at 7 p.m. and arrived in Avery at 4 a.m./*Photo by Gayle Christen.

F Engine 6209, a 4-6-2 waits in Spokane on 6-6-38. This locomotive later became the 889 and was the last F Engine to run on the Idaho Division./Photo by Albert Farrow.

The F Engines were basically light passenger locomotives. Their power and weight was typical for 4-6-2's built around 1910. They operated at a boiler pressure of 200 lbs. per square inch, they could easily make 60 to 70 mph with a train, and they could haul 8-10 coaches up a 1.7% grade.

System-wide, passenger service was the F Engines' primary work. Freight was assigned to the more powerful L class 2-8-2's. On the Idaho Division however, things were different. No 2-8-2's were based there and the F Engines were regularly used in freight service. A few F Engines were used exclusively for freight and did not even carry steam lines for passenger train heat. The reasons for this type of motive power utilization are explained through the memories of retired employees:

> "We used the F Engines on branchlines that didn't have heavy rail," says retired engineer Pat Chester, and he adds, "When a branchline engine got on the mainline, it had to be able to outrun the passenger trains. The L Engines were pulling engines but they couldn't make speed."

As retired engineer Paul Edmunds put it:

> "The F Engines were faster and easier on the track than the L Engines. The L Engines were very rigid and you couldn't take them into some of the industry tracks."

In addition to running in general freight service, the F Engines were the normally assigned helpers on Mica Hill. The Hill was a seventeen mile 1.7% grade coming up from the Spokane Valley. The helpers assisted both freight and passenger trains.

The general purpose F Engines remained in service until the early 1950's when they were replaced by general purpose EMD hood units. The first to be scrapped was the 847, which left the roster in January of 1951. The last was the 889, which was scrapped in December of 1954. The 889 is remembered for its smooth ride and easy steaming abilities.

THE L ENGINES
CLASS L-2, 2-8-2's

The largest group of steam locomotives on the Milwaukee Road were the 2-8-2's. There were 437 of them in three classes and they were the standard freight engine for most of the system. On the Rocky Mountain Division and the Coast Division, L Engines served as branchline power but none were assigned to the Idaho Division.

Several Coast Division L-2's were based at the division point of Othello. Their regular assignment was the Hanford branch, a line which was 106 miles closer to Othello than Tacoma. If traffic warranted, the L Engines were also used on Idaho Division runs. Their assignments were mostly freight but some could be used in passenger service. Engines 711 and 733 had steam lines for train heat and were occasionally used to pull #7 and 8, the Butte Local between Avery and Spokane. During those times when the L Engines were in Spokane, they also served as helpers on Mica Hill.

The L-2 class 2-8-2's were built by ALCO from 1912 to 1914. They rode on 63 inch drivers, operated at a boiler pressure of 200 lbs., and had a tractive effort of 54,723 lb. The statistics are almost identical to a U.S.R.A. Light Mikado.

On the Idaho and Coast Divisions, the L Engines are widely remembered as hard riding, slow, powerful engines that used large amounts of water. Some crews on the Rocky Mountain Division nicknamed them "Stump Pullers."

L Engine #665, a light 2-8-2, takes water at Spirit Lake, Idaho, on a cold winter day. Date unknown./ Photo from Jack Powers Collection.

CLASSES C, I-5, AND K, 2-8-0's, 0-6-0's AND 2-6-2's

The C Engines were 2-8-0's built between 1910 and 1912. Some were constructed by the Milwaukee shops and others by ALCO. There were seven on the Idaho Division and they belonged to classes C-2sr, C-5 and C-5a. Their drivers were 63 inches in diameter, boiler pressure on some was 185 lbs. and on others 200 lbs. Tractive effort was roughly 43,000 lbs. for all of them. Their normal assignment was the Moses Lake branch, the St. Maries branch from St. Maries to Bovill and Elk River, work trains, and general yard switching chores. The first of the Idaho C Engines was retired in 1951 and the last in 1954.

The I-5's were a group of five 0-6-0's used for switching in Spokane. They were built from 1903 to 1912 by the Milwaukee shops. Their drivers were 51 inches in diameter, boiler pressure was 180 lbs., and tractive effort was 28,160 lbs. The I-5's were easy steaming engines and were often used for breaking in new firemen on the Idaho Division. The first of the five was retired in 1951, the last in 1954.

The K Engines were 2-6-2's, a group of hard riding locomotives considered too slow for the Idaho Division. None were assigned there, but one, the 924, was based in Othello. It was assigned to the Coast Division's Hanford branch. The 924 had 63 inch drivers, carried 200 lb. boiler pressure, and had a tractive effort of 33,320 lbs. It was built by ALCO in 1908 and retired in 1955. It was not based in Othello when retired.

#1405, a Class I-5 Switcher at Spokane. Date unknown. According to Bill Plybon, "The I-5's were stout and had big cylinders. We used to pull some awful loads with them."/Photo by Gayle Christen.

#1201 was built in the Milwaukee, Wisconsin shops in April of 1912. It operated until the end of Idaho Division Steam in December of 1954. Othello, date unknown./ Photo by Gayle Christen.

*K Engine #924 sits by the roundhouse on one of its infrequent visits to Avery. Date unknown./*Photo by Gayle Christen.

I-5 yard goat #1481 switches the long narrow yard at Spokane, Washington on March 10, 1951./Photo by Philip R. Hastings.

High stepping Speedster #132 wheels the eastbound Columbian down the Union Pacific line that Milwaukee crews called "The Freeway." The location is Marshall, Washington and the date is August, 1950./Photo by Philip R. Hastings.

131 AND 132
CLASS F-6, 4-6-4's

The Mechanical Department in Milwaukee labeled these 4-6-4's the Baltics and Class F-6. Other roads called them Hudsons, but none of those names were used on the Idaho Division. The two engines assigned to The Gap were simply known as 131 and 132. They weren't called F Engines in order to avoid confusion with the 4-6-2's.

The Milwaukee was the first railroad to design a 4-6-4, but due to financial problems, none were built until 1930. By that time the wheel arrangement was already in service on the New York Central, the Santa Fe, and the Canadian Pacific. The New York Central's Hudsons comprised almost two-thirds of the 4-6-4's built in North America, thus the type became most closely associated with the N.Y.C. In the areas of performance, however, the Milwaukee's Baltics were not outshined.

The F-6 class was a group of high stepping speedsters designed to improve the Milwaukee's passenger schedules in the Midwest. It is a job they did very well. Most were used on the high speed route between Chicago and Milwaukee where, in 1934, the 6402 established a world speed record. Pulling a five car train, it achieved a sustained average speed of 92.62 mph for a distance of 61.4 miles. Its peak speed on the trip was 103 mph, and the terminal to terminal average for the 86 mile trip was 76.07 mph. Later in the year, the 6415 demonstrated how rugged and reliable the Baltics were. Operating on the 914 mile Minneapolis to Harlowton run it made ten round trips in 30 days for a total of 18,390 miles, without requiring shop work.

The 14 original Baltics were so successful that the Milwaukee bought eight more in 1931. The second group was classed F-6a. Both groups had 79 inch drivers, 225 lbs. of boiler pressure and 45,250 lbs. of tractive effort at 25% adhesion. They were built by Baldwin Locomotive Works. The original road numbers were 6400 to 6421. After renumbering in 1938, they became 125 to 146. The larger, faster F-7 streamlined Hiawatha 4-6-4's which the Milwaukee bought in 1938 were numbered 100 to 105.

Almost from the time the Baltics entered service in 1930, there was speculation that these locomotives were just what the Milwaukee needed on the Idaho Division. None were assigned to The Gap until 1944 and 1945 when the 131 and 132 were converted to oil and sent west. Although they were faster than the F Engines, 131 and 132 could not pull a significantly heavier train. The 4-6-4's could only pull a 9-10 car passenger train on the Idaho Division. The 4-6-2's could pull 8-10 cars. In freight service, the 131 and 132 were rated at roughly 75% of the tonnage handled by the 4-8-4's. Low tractive effort kept their train length down, while high engine weight kept them off of the branchlines where the F Engines were so successful.

Due to limited utility, the Milwaukee sent no more Baltics west. The only railroad in the Pacific Northwest which truly considered 4-6-4's successful was the Canadian Pacific. That railroad placed five Royal Hudsons in service from Vancouver B.C. to Revelstoke.

Bill Plybon, whose career stretched from 1912 to 1965 remembers the Idaho Division 4-6-4's very well.

> "The 131 and 132 were a real light, high speed, passenger engine they had back east. They had no business sending them to the Idaho Division," he says. Plybon was, however, impressed with their speed and he recalls: "One time I was running on the U.P. between Spokane and Marengo with the 131 or 132 when my fireman said to me, 'I wonder how fast this engine will go?' I said to him, 'I don't know, but we'll find out.' We were on a long straight stretch when I opened up the throttle. We got

up to 94 mph and then I had to ease back. There was a curve coming up and we were running out of railroad.''

Pat Chester, who fired the 131 and 132 remembers them as:

''Very nice engines that would get right out and move. They were good steamers and had a smooth ride,'' he says.

When the Milwaukee began dieselization in earnest on the Idaho Division, the light, fast 131 and 132 were the first modern steam engines to go. They were retired in 1952 and towed to Seattle for scrap.

#6400 first of the class called Baltics by the Mechanical Department. It is shown at Minneapolis, 5-9-37./Photo by Graham.

The 131 is shown backing near the Othello depot on 7-21-52. ''The 131 and 132 rode good and steamed good,'' according to Paul Edmunds. ''We used to run them 75 mph and better down by Roxboro and across the flats.''/Photo by Wally Swanson.

ORPHAN ANNIE AND THE S ENGINES
CLASSES S-1 AND S-3, 4-8-4's

The 9700 was Orphan Annie. It was the first 4-8-4 on the Milwaukee and the only one running during the first seven years of the depression. It was introduced into service at a time when America was introduced to its namesake little girl in the comic strips. This lone 4-8-4 was like an orphan and the nickname stuck for many years.

The Milwaukee emerged from bankruptcy as the depression began. Though most of the country was in a deepening financial crisis, the Milwaukee's finances were in good shape. Thus it upgraded passenger service and built the locomotives it could not afford in the mid 1920's.

The 9700 was a 223 ton locomotive which measured 101 feet from pilot to tender drawbar. It rode on 74 inch drivers, carried 230 lbs. of boiler pressure and produced 62,136 lbs. of tractive effort at 25% adhesion.

It was built by Baldwin Locomotive Works in March of 1930 and assigned the class name S-1. It was part of a group of fifteen locomotives which Baldwin built for the Milwaukee that year and it was something of an experiment. The Milwaukee wanted to know if a 4-8-4 would outperform a 4-6-4 in passenger service. Fourteen 4-6-4's and one 4-8-4 were built and compared.

Orphan Annie was initially assigned to the 914 mile Minneapolis to Harlowton run along with a class F-6 4-6-4. When the 4-6-4 was found to be quite capable of pulling the Olympian across the Prairies, the Milwaukee decided to build more 4-6-4's. Eight more were constructed in 1931 and Orphan Annie was assigned to freight service. It worked out of Chicago's Bensenville yard until approximately 1933, when it was converted to oil and sent to the Idaho Division. Its assignment was pulling the Olympian across The Gap.

According to long time engineer Bill Plybon:

> "The 9700 was our first big passenger engine. It was really nice and as soon as it got here everyone wanted to run it."

Orphan Annie was very successful in Idaho Division service. As a result, the Milwaukee shops built a duplicate 4-8-4 in 1938. The new locomotive was also assigned to The Gap. Due to a general renumbering of steam locomotives, the new locomotive entered service as the 251. Orphan Annie, the old 9700, became the 250.

These locomotives had the power to pull heavy trains and roll them fast. With bigger boilers and taller drivers than the older locomotives, they made a dramatic improvement in Idaho Division motive power. They could easily roll passenger or freight trains at 80 mph, if a fast running engineer opened them up. Most assignments were in passenger service until diesels arrived in numbers. After that, they were used in passenger or freight according to the whims of the dispatchers.

According to Paul Edmunds, who began firing on the Idaho Division in 1943, the 251 was the better steaming engine of the two.

> "It had a little more pep to it and it would talk to you, going up the canyon. On the 250, you had to have the oil hotter. It was a hard steamer."

In addition to keeping the steam pressure up, train crews were concerned about keeping the smoke up. When the engines were drifting downgrade with the throttle almost closed there was little exhaust steam to blast the smoke high out of the stack. As a result, smoke swirled low around the engine and tended to obscure the engineers' view of switches and signals. In

"Orphan Annie," #9700 sits in a yard somewhere in the Midwest with her bell ringing and a full load of coal in her tender. She was converted to oil before going west to the gap. Date of photo is unknown./Photo from Harold K. Vollrath Collection.

251 at Spokane's Union Station 7-15-38. A sight to make a hoghead proud. This S-1 class 4-8-4 was built in the Milwaukee, Wisconsin, shops./Photo by Albert Farrow.

Orphan Annie as she looked in July of 1950. She has a new number and a fold-down stack extension ahead of her smokestack. Her bell and headlight also have new locations. She is awaiting a train at Othello, Washington./Photo from Harold K. Vollrath Collection.

order to remedy this problem, the Milwaukee applied mechanically raised stack extensions to both engines. This did little good according to Paul Edmunds who says:

> "The best way to lift the smoke was to set a little air on the train brakes and make the engine work."

The Milwaukee's second group of 4-8-4's were the S-2's. These were forty burly Baldwin built freight haulers which were delivered from 1937 to 1940. They worked the Midwestern mainlines east of Harlowton. None ever ran on the Idaho Division. If things had been different, they might have come west, but in 1940, General Motors FT diesel demonstrators took the Idaho Division by storm. From 1941 through 1944, the Milwaukee bought four 4-unit FT sets for the Idaho Division. It also bought four sets to use elsewhere and it wanted more. An attempt to purchase more diesels was made at the height of World War II, but the War Production Board allocated steam engines instead. The ten 4-8-4's which were received became class S-3.

The S-3's were built by ALCO in 1944 and numbered 260 to 269. Due to wartime restrictions they were a compromise locomotive. ALCO took a 1929 Rock Island design, modernized it somewhat and built variations of it for the Milwaukee, the Rock Island, and the Delaware and Hudson. The Milwaukee version had roller bearings, an enclosed cab, and a water bottom tender.

The S-3's rode on 74 inch drivers, weighed 230 tons, and carried 250 lbs. of boiler pressure. Their tractive effort was 62,119 lbs. at 25% adhesion. This was the lowest of the three S classes, but it was close enough to an S-1 that Idaho Division dispatchers equated it with an S-1 for tonnage ratings.

The S-3's worked exclusively in the Midwest until the Korean War brought an acute power shortage to the Idaho Division. In 1950 the 262, 263, 267 and 269 were taken from the D&I Division in Illinois, converted to oil one at a time and sent west. On the Idaho Division they were well received because they were easy steaming, smooth riding engines, and they proved to have significantly more pulling power than the S-1's. Crews also appreciated the enclosed cab which kept them warm in the wintertime.

The S-3's and S-1's remained on The Gap until December of 1954. That month was curtains for all Idaho Division steam. Dieselization was complete. The 267 made steam's final run on December 17th.

Following retirement, the S-3's were sent to Tacoma and stored. They were not yet paid off, and the Milwaukee could not scrap them until the mortgage note was retired. None of the Idaho Division steam engines were needed elsewhere on the system and all classes were eventually scrapped. The only representatives of Idaho Division steam which remain are two S-3 class 4-8-4's which never ran on the Idaho Division. The 261 is on display at the National Railway Museum in Green Bay, Wisconsin, and the 265 is displayed at the Illinois Railway Museum in Union, Illinois.

S-3 engines like the 269 were smooth riding and easy to fire. They were also "strong on the tonnage rating." Because of this, the Idaho Division crews loved them. The scene is Spokane in 1953./Photo by Jim Fredrickson.

Train #18, the Columbian, leaves Spokane, Washington on April 20, 1952. The 262 and eight cars are eastbound./Photo by W.R. McGee.

Train #263 arrives in Harlowton, behind S-2 engine 229 on April 23, 1941. The S-2's were the largest, most powerful steam engines on the system. Fourteen were assigned to the Minneapolis to Harlowton run. None ever operated on the Gap./Photo by W.R. McGee.

DIESELIZATION

The first step toward dieselization of the Idaho Division came in 1940. It came in the form of a four-unit FT demonstrator. Electro-Motive Division of General Motors built a group of FT's in 1939 and dispatched them on a barnstorming tour of the American rails. Offered to the railroads free of charge, they were pitted against steam in head to head competition to demonstrate their worth. They did just that and in 1940 the set which ran on the Idaho Division ran the pants off of steam. As a result, the Milwaukee bought its first road diesel in 1941. This four-unit locomotive was numbered 40ABCD. It was rated at 5,400 hp, 220,000 lbs. of tractive effort, and was equipped with dynamic brakes. It also carried a train heat boiler in order to allow its use in passenger service.

The 40 quickly impressed Milwaukee Road officials for a number of reasons. Due to the high tractive effort available in a four-unit locomotive, it could pull 70% more tonnage than a 4-8-4 from Othello to Avery. On the 1% westbound grade from Avery to Ramsdell it could pull 140% more tonnage than a 4-8-4. Steam could match the tonnage if engines were double or triple-headed, but that meant paying several crews instead of just one. Because the 40 did not require extensive servicing in terminals it could make three trips per day across the division, as opposed to one for a steam engine. Over the road times were better because the 40 pulled so well on grades and it did not have to make water stops. Milwaukee Road officials were impressed and they quickly made efforts to get the most out of their new road diesel. They regularly assigned it demanding tasks and cleared the track for its runs. When time

Famous diesel #40 in 1941 builder's photo. "The 40 had dynamic brakes, but I never liked using them," says Bill Plybon. "You could only go down Plummer Hill at 15 mph if you used the dynamics."/Photo courtesy EMD.

On the Avery ready tracks, are Freight Motor E57 AB, Mallet #53 and brand new diesel #40. Date and photographer unknown.

orders were issued to steam engine crews, they bore a cryptic message in large type across the bottom, "DO NOT DELAY THE 40."

The Milwaukee had long looked forward to electrification of the Idaho Division, but the money was never available. With the arrival of the 40, however, the opportunity seemed at hand. Modern technology offered DIESEL-electrification. Diesels could be purchased a few at a time and phased into service over a period of years. This made them easier to acquire than substations, trolley wires, and electric locomotives. The latter had to be purchased all at once to be useful. In addition diesels were seen by the Mechanical Department as economically superior to electrification.

Dieselization progressed slowly at first. Three more four-unit FT sets were assigned to the Idaho Division during World War II. Then the Milwaukee's attention shifted to other divisions and to the dieselization of Hiawatha passenger service. The FT's worked alongside steam and, because of their train heat boilers, it was not uncommon to find them pulling heavy troop trains.

In 1947, the Milwaukee replaced the old Olympian passenger train with a new lightweight streamliner called the Olympian Hiawatha. The new train was pulled from Chicago to Tacoma and back by Fairbanks-Morse diesels. The majestic, but trouble-prone, silver nosed "Erie-Builts" ran straight through without change until 1949. In that year, a nationwide coal strike idled many of the Milwaukee's steam locomotives. As a result the Fairbanks-Morse diesels were held in the Midwest to supplement the use of coal burners. They were never again used west of Harlowton.

Erie-built Fairbanks-Morse diesels and the Olympian Hiawatha were the Milwaukee's flagship when they entered service in 1947.

The luxurious Olympian Hiawatha had just recently been inaugurated when these photos were taken at Spokane's Union Station. The eastbound train arrived at 10:55 p.m., while its westbound sister arrived at 12:01 a.m./Photos by Bud Tilbury.

Train #261 near St. Maries, Washington, March 1964./Photo by Richard Steinheimer.

In 1949, two F7 freight diesel sets were assigned to the Idaho Division, but no more diesels arrived until 1954. In that year, 32 GP9's were received and steam was promptly retired. Five of the GP9's were equipped with train heat boilers in order to serve as passenger locomotives. The next group of diesels came in 1959. They were 50 additional GP9's. Among other things, this group replaced the four sets of FT's. The worn out FT's were 15 to 18 years old and became trade-ins.

Passenger diesels such as E7's and FP7's were seen on the Idaho Division beginning in the late 1950's when they ran through from Chicago with the Olympian Hiawatha. They disappeared in 1961 when passenger service ended.

Beginning in 1965, the Milwaukee purchased a fleet of new larger locomotives to replace the GP9's. The locomotives came from both General Electric and EMD. The purchases stretched into the 1970's. During this time, the utilization of motive power also changed. Some locomotives ran through from Chicago to Tacoma, some from Avery to Tacoma and others simply ran on The Gap.

Dieselization brought major changes for the engine crews and several are willing to share their memories.

THE EARLY DIESELS

"When the diesels first came out I thought they were great" says Pat Chester. "They were warm and clean. I caught a cold every winter on steam. We had diesel riders from the factory to show us how to run them so it wasn't hard to learn, but some of the old-timers fought diesels. They liked steam better because running diesels was like running streetcars to them. Steam got pretty bad towards the end though. Once the Milwaukee started getting diesels they abused the steam something fierce. Some things would wear out or break and never get fixed and they would cannibalize one engine to keep another one running."

"I fired on the first diesels we had, the 40 and the 41. They had manually controlled radiator shutters. If a fireman got careless and let the radiators get too hot he could do a lot of damage to the engine."

"One selling point of the diesels when they first came out was that nobody needed to take care of them. With steam, you had to hook it up to a steam line in the roundhouse to keep it hot all night. They said you could just turn off a diesel at night and then start it up again the next morning. Well, it didn't work that way. It takes a long time for those big engine blocks to warm up, and if the locomotive was run before the block was warm you would tear up the cylinder liners and break the cylinder heads. They had to run the diesels all night to keep the block warm. Diesels weren't supposed to break down either, so if you saw them working on one in the shop, the repair ticket might say "Rod work for mallet #55." After the steam engines were gone, the Milwaukee would charge off diesel repairs to the electrics. They didn't want the diesels to look bad."

"After diesels came out, there was a lot of propaganda that firemen weren't needed anymore. The people who said it didn't run trains though. If you have a consist of several units, someone needs to patrol them regularly to make sure that they don't run low on oil or water and burn up. If they stop, the fireman has to start them, and if they don't want to run, the fireman has to nurse them along so the engineer can get the train into the terminal. The first diesels weren't as automatic as the salesmen said they were, and neither are the new ones."

THE FAIRBANKS-MORSE DIESELS

"There were a lot of problems with the injectors and the load regulators on the Fairbanks-Morse," says Paul Edmunds. "After we finally got the kinks out of the Fairbanks-Morse, they started having mechanical problems. They just didn't work out well here. I don't know if it was because they didn't get the maintenance they needed or because they weren't made for this kind of territory."

THE DIESELS BUILT BY ELECTRO-MOTIVE DIVISION OF GENERAL MOTORS

These locomotives were universally considered to be the best by Milwaukee crews. Because of excellence in design and construction they became the standard of comparison for other brands and types of locomotives.

THE GENERAL ELECTRIC DIESELS

"The GE diesels were terrible," says King Clover, who ran them on the Coast Division. "On the older ones you would have to anticipate everything you wanted to do. You would haul out the throttle and it would just sit there going ker-thump, ker-thump until it finally revved up and it might take a mile to do that. Also, they were dirty, oily and the doors wouldn't stay shut. The 8000's (U33C's) were good engines, but at high speeds they would move laterally so bad they would knock you right out of your seat."

According to Paul Edmunds, "The people who designed the GE diesels must have figured they would never have to work on them. The GE diesels used hot air off the manifold to heat the cab and you got a lot of gas in the cab if the engine had a leaky manifold. The GM diesels had a good hot water heater in the cab. The GE diesels had poorly arranged controls too. They had a long throttle arm and a short air brake arm. You had trouble using both of them at once, and if you were looking out the window you couldn't reach the brake. If you had to start them or reset the overspeed relay you had to come out of the cab, go around the back and all the way up the other side to get to the cabinet where the switches were. On a GM diesel that cabinet is right outside the cab door."

Pedee Bridge west of St. Maries, Idaho, September 4, 1976./Photo by Jim Fredrickson.

"The Milwaukee had a lot of arguments with General Electric," recalls Wade Stevenson, who worked in the Othello shop. "Those diesels constantly had trouble with oil leaks, and water leaks. The Milwaukee said it was because of poor design and workmanship. GE said it was because the Milwaukee wasn't doing any maintenance. Whatever the problem was, the GE diesels didn't run as well as the GM diesels did."

The Milwaukee's response to its problems with GE diesels was to withdraw them from the Pacific Northwest in 1977. Only a few of them returned in 1980. Idaho Division crews have no memories of ALCO diesels because they were never sent west. From beginning to end, the story of diesels was largely a story made by General Motors.

F7 diesel #88 ABC crosses Lake Chatcolet trestle with a freight train in 1955./Photo by Jim Fredrickson.

Diesel railroading in The Gap...#261 skirts Rosalia, Washington on a wooden trestle as three SD40-2's work west for Malden...it's September, 1974, and with the demise of electric operations, the entire division is due for changes in operating procedures. September 1974./Photo by Ted Benson.

Just above Royal City and in the shadow of the Smyrna cliffs, a local freight waddles past Taunton. 8-28-78./Photo by Blair Kooistra.

Switching the yard at St. Maries, Idaho, in 1968./Photo by Jim Fredrickson.

St. Maries depot in 1968./Photo by Jim Fredrickson.

Malden, Washington on September 22, 1974—#262 eases to a stop outside the depot for a crew change at a quarter past three on a pleasant Sunday afternoon. Lead SD40-2 176 (one of three) sports a single bell cowhorn that could well have come off a recently-retired Joe. Chasing this train east gave some eerie sensations in the aftermath of May's final electric photos./Photo by Ted Benson.

As the September sun settles into the 5 o'clock shadows, the conductor boards the caboose of First #261 at Malden, making one of his last trips to Othello before both Malden and Avery were sacrificed in favor of a new crew terminal at St. Maries. I could have never known it then, but this was the last shot I ever made of the Milwaukee Road in the West. Even if it wasn't electric, it wasn't too bad! September 22, 1974./Photo by Ted Benson.

LOCOMOTIVES REGULARLY OPERATED ON THE IDAHO DIVISION

STEAM LOCOMOTIVES

CLASS	TYPE	ROAD NUMBERS
N-3	2-6-6-2	51, 52, 53, 54, 55, 56, 57, 58, 59, 64, 65
S-1	4-8-4	250, 251
S-2	4-8-4	None
S-3	4-8-4	262, 263, 267, 269
F-6	4-6-4	131, 132
F-5b	4-6-2	835, 847, 848, 849, 851, 853, 855
F-4ms	4-6-2	875
F-4b	4-6-2	889, 890
L-2	2-8-2	*665, 684, 687, 688, 711, 712, 726, 733
K-1	2-6-2	*924
C-5	2-8-0	1201
C-5a	2-8-0	1211, 1213, 1220, 1222, 1229
C-2sr	2-8-0	1262
I-5a	0-6-0	1405, 1459, 1481, 1489, 1491 (Spokane switch engines)

*These were Coast Division locomotives and they primarily worked the Hanford branch which is on the Coast Division. They were based in Othello because Hanford was 106 miles closer to Othello than to Tacoma.

NOTE: •All steam engines on the Idaho Division were oil burners.

•The last run of a steam engine on the Idaho Division was by #267 on December 17, 1954.

•Roster data courtesy of Wade Stevenson.

LOCOMOTIVES REGULARLY OPERATED ON THE IDAHO DIVISION

DIESEL LOCOMOTIVES

Make	Model	HP Per Unit	Comments	Road Numbers
EMD	FT	1,350		40ABCD, 41ABCD, 42ABCD, 43ABCD
EMD	E7	2,000		16 AB
FM	Erie	2,000	Chrome nose	5-9 ABC, 10 AB
FM	Erie	2,000	Painted nose	21 ABC, 22 AB
EMD	F7	1,500		87 ABC, 88 ABC
EMD	TR2	1,000	Branch lines	2000 AB
EMD	TR4	1,200	Branch lines	2001-2006 AB
EMD	FP7	1,500		104 ABC, 105 ABC
EMD	GP9	1,750	Dynamic brake equipped	280-331
EMD	GP9	1,750	No dynamic brakes	2400-2425
EMD	GP9	1,750	Steam boilers, no dynamics	2426-2431
EMD	E9	2,400		30 ABC, 35 ABC
EMD	GP30	2,250	Tacoma-Bensenville	340-355
EMD	GP35	2,500	Tacoma-Bensenville	360-371
EMD	GP40	3,000	Tacoma-Bensenville	2000-2071
EMD	SD40-2	3,000	Tacoma-Bensenville	16-30, 130-176, 182-209
EMD	SD45	3,600	Tacoma-Bensenville	4000-4009
EMD	FP45	3,600	Tacoma-Bensenville	1-5
GE	U23B	2,250	Branch lines	4800-4804
GE	U25B	2,500		5000-5010
GE	U28B	2,800		5500-5511
GE	U30B	3,000		6000-6009
GE	U33C	3,300		8000-8003
GE	U36C	3,600		8500-8503

NOTES:
1. Not all of these locomotives were on the Idaho Division at any one time.
2. Some locomotives may have been omitted due to lack of information.
3. Many locomotives were renumbered during their careers.
4. Passenger locomotives such as E7, FP7, and Erie ran through. They were not assigned to the division. Locomotives marked Tacoma-Bensenville ran through on freight.

Roster data courtesy Wade Stevenson and Don Dietrich.

ECONOMICS AND COMPANY POLICY

The economic benefits which could result from electrification flowed to many companies in addition to the Milwaukee. The project was potentially a source of revenues for Anaconda Copper Co., General Electric, Westinghouse, Great Falls Power, Thompson Falls Power, Washington Water Power, and Puget Sound Power & Light. All lobbied heavily to promote the project. Most active among the lobbyists was John D. Ryan, who was elected to the Milwaukee's Board of Directors in 1909. Ryan was president of Anaconda Copper and a board member for both of the Montana based utilities which later merged to become Montana Power.

When Ryan and others convinced the Milwaukee to electrify, the result was a number of profitable business deals occuring between 1915 and 1920. These included the sale of 11,000 tons of copper wire, 103 locomotives, 46 transformers and 46 motor-generator sets. It also meant the signing of lucrative 99 year power supply contracts which were personally negotiated by Ryan. These contracts gave his utilities their first large customer. The Milwaukee's 654 mile electrification was uniquely important to General Electric and Westinghouse because it demonstrated the practicability of long distance steam railroad electrification.

Previous electrification projects were all under 160 miles in length and many were built just to allow operation in poorly ventilated tunnels, or to haul commuters and to comply with urban smoke abatement laws.

The Milwaukee's announced reasons for electrifying were to improve general operation and to reduce operating costs. It was successful in both areas. Electrics easily bested the schedules operated by steam, and they did so at an operating cost which was only 54% of what it would cost to run steam. Savings were not just from speed and fuel economy. They were also due to the use of fewer locomotives, fewer engine crews, less required maintenance, fewer roundhouses and a smaller maintenance staff.

Although these savings were substantial, they did not rock the railroad world or silence doubters in the Milwaukee's own management. These substantial savings were achieved at substantial expense. When the construction work was completed in 1920, the bill for substations, trolley and a fleet of locomotives came to $23 million. That was three times the amount previously spent to equip the same line for steam operation. This large investment exacted a heavy toll as bond interest and payments devoured much of the money which was saved. Additionally, the bond indebtedness pushed the Milwaukee closer to bankruptcy.

According to a report published by the Milwaukee in 1925, the net annual cost reduction on the Rocky Mountain Division averaged $1 million after bond and interest payments. Savings on the Coast Division were less impressive, averaging only $100,000 per year. The cause of the Coast's poor performance was inflation. Due to rising prices after World War I, the per-mile cost of electrifying the Coast Division was twice as high as the Rocky Mountain Division. Interest rates had increased also and the resultant payments used up most of the electric versus steam cost differential. Although the economics would improve if traffic grew, no growth occurred.

In the view of electrification's critics, both inside and outside of the Milwaukee's management, the company spent a great deal of money to achieve a small change in the financial bottom line. The American railroad industry was for the most part committed to the use of steam, and it tended not to seriously consider other types of motive power for rail lines where

Freight Motor E64 AB at Tacoma Yard 9-9-50./Photo by Albert Farrow.

Gold Creek, Montana, is better known as the site of NP's last spike—duly recognized at a 1983 centennial observance—but 11 summers earlier, the real precious metal in town was the steel and copper of the E50 grinding east for Deer Lodge with train #266. Neither E50 nor the Milwaukee was on hand for the big party in 1983. July 17, 1972./Photo by Ted Benson.

*Although #80 ABCD was geared low for branchline log train service, it proved to have a slightly lower mainline operating cost than three-unit GE Motors in 1948. The diesels were new and the electrics were 33 years old./*Photo courtesy EMD.

steam was adequate. Railroads generally rejected electrification as unnecessary or too expensive. During the steam era, further investment in the Milwaukee's electrification was stifled by a number of factors in addition to skepticism. A bankruptcy lasting from 1925 to 1927 was followed by the Depression and reduced business. That in turn brought on a second bankruptcy which lasted from 1935 to 1945, a period which included World War II.

The advent of mass produced diesels brought challenges to the existence of the electrification. Because they possessed many of the same advantages possessed by electrics, their cost of operation was much closer to that of electrics. The Milwaukee debated for years about which was the better investment.

A study performed by Laurence Wylie in 1949 indicated that new diesels would be slightly cheaper to operate on the Coast Division than the old electrics. The purchase price, interest and taxes are what made diesels a more expensive choice.

This closeness of costs was not unique to Milwaukee. A 1948 study on the Great Northern electrification found diesels to be slightly cheaper to operate there also. The Great Northern blamed this on the exorbitant rates it was paying for electric power. Its variable rates went as high as $.01 per kilowatt hour (kwh) including demand charges. Joe Gaynor, who headed the G.N. electrification, felt the rate would have to be cut to $.005 per kwh for economics to justify expansion of their electrification.

Energy costs were one of the factors favoring electrics on the Milwaukee. From the late 1940's through the late 1960's, the Milwaukee paid $.09 per gallon for diesel oil compared to $.068 for enough electricity to produce equivalent power. (Calculations and locomotive tests showed that one gallon of oil equaled 12.5 kwh which were purchased for $.00544 per kilowatt hour.) This rate was quite good in comparison to the Pennsylvania RR which paid roughly $.009 in 1950 and $.013 by 1966. Bonneville Power Administration once offered electricity to the Milwaukee for $.004 per kwh, but the B.P.A.'s minimum use charge was built around more power than the Milwaukee could consume. For that reason, the B.P.A. rate was actually higher than the amount the Milwaukee paid to private utilities.

*Train #263 last westbound electric negotiates Jefferson River Canyon on June 14, 1974. Power: E79, E71, 3013, 3024 and 296./*Photo by Dick Dorn.

Diesels west...first section of 261 has Bitterroot crossing all but whipped as a trio of SD40-2's pass the abandoned East Portal depot on a snowy Mother's Day afternoon. May 1974./Photo by Ted Benson.

A major factor in the Milwaukee's diesel versus electric cost studies was locomotive repair costs. According to Electrification Department estimates, these favored electrics by a substantial margin. Unfortunately, the company was never able to verify that. The Mechanical Department was responsible for maintenance to both types of power and its records were considered unreliable. It was widely accused of padding the costs assigned to electrics and hiding costs attributable to diesels. Comparative studies thus had to include several estimated maintenance costs in order to show the economic outcome if different estimates were true.

In 1968, Laurence Wylie, who was retired and working as a consultant, prepared one of the most thorough studies the company reviewed. Among other things, he analyzed what the results would have been if the railroad had electrified The Gap and purchased 29 new 5,000

Crippen held back on deciding the fate of the electrification and in the meantime the electric fleet dwindled far beyond the point of cost effective operation. Crippen, however, may have been distracted by the company's larger problems. The Milwaukee was sliding toward its third corporate bankruptcy and the economic picture looked gloomy.

The Milwaukee's net income for all of 1966 was $14.7 million. It dropped to $5.1 million in 1967 and rose to only $5.2 milion in 1968. That was the company's last profitable year. The national economy went into recession, and this was followed by simultaneous inflation and recession. Railroads everywhere suffered, but the Milwaukee fared worse than most. Operating revenues in 1969 stood at $275.6 million, but revenues were exceeded by costs. The Milwaukee netted a $5.6 million loss. Losses grew to $11 million in 1970 and $21.8 million in 1971.

Crippen moved out and up to become Vice-Chairman of the Board in 1971 and was succeeded as President by Worthington Smith. Smith felt the electrification issue should be resolved, and shortly after taking office, he appointed an 18-member committee to make a recommendation. The group included the Vice-President for Operations, the Vice-President for Finance, the Vice-President for Planning, the Chief Purchasing Officer, the Chief Mechanical Officer and the Electrical Engineer among others.

As the group debated the choices before it, some members questioned the Milwaukee's ability to pay for any major new investments. Many also took note of the negative impacts

Milwaukee's biggest locomotives on the west end in later years were the group of General Electric U36C's 3600 horsepower of chuff-chuff locomotion, referred to by some railroaders as "Toasters" due to GE's experience in manufacture of household appliances. Units 5802 and 5803 were kept on the west end for the last couple of years of operation, usually assigned on Tacoma—Othello runs. It was a rare occasion then, to catch BOTH of the U36C's on an eastbound extra, in company with recently assigned GP40's 2019/2036, seen here blowing snow at more than the allowable speed limit at Thorp, Washington on February 16, 1980... one day short of a month before the absolute end./Photo by Blair Kooistra.

The "last" train #2017, this is officially #200-S-1, the last eastbound revenue freight train on the Milwaukee out of Tacoma. It's March 1, 1980 and the embargo of service is in effect. Here GP40 2017 leads four sisters and a rare (on the Washington Division) GP30 off the Columbia River Bridge and into what's left of the town of Beverly./Photo by Blair Kooistra.

which could occur if the Milwaukee abandoned some of its Montana trackage and ran on the Burlington Northern, or if the two companies merged and traffic was diverted. The committee recommended against retaining the electrification. Smith concurred with them and so did the Board of Directors. The decision was announced in February of 1973.

Losses were continuing to mount on the Milwaukee, and in 1973 the company found it difficult to make bond and interest payments. The future looked bleak due to the company's inability to compete with the much stronger Burlington Northern. Opting to join them rather than fight them, the Milwaukee's directors proposed a merger. A year and one-half later, when the B.N. rejected the proposal, the Milwaukee had already been sued by bond holders seeking overdue payments.

The mood in the Milwaukee's Chicago headquarters was one of deep depression. The next three years brought frantic requests that the Interstate Commerce Commission restrain the B.N.'s competition or force the B.N. to merge with the Milwaukee. The I.C.C. declined to do either, but the Milwaukee did acquire Federal loans and it made plans to ask for more of them. The Federal government seldom let railroads liquidate assets to pay their debts, it helped them back on to their feet with loans and protective regulations. The Milwaukee had been bankrupt twice before, but through corporate reorganization it had recovered. Many felt it would do that again. Outside of the Milwaukee's highest officers, few employees knew or

Whining downgrade at a steady 15 mph through Washington's Cascade Mountains, westbound train #201, with SD40-2's 22/164/167/26 strides over the curving steel bridge at Change Creek, above the old siding at Ranger, Washington at 5:41 p.m. May 25, 1979./Photo by Blair Kooistra.

Striking a grand pose seemingly defiant of the financial troubles that will consume the Pacific Extension nine months later, Milwaukee Road train #200 blasts up the 1.74% St. Paul Pass grade above Falcon, Idaho on June 11, 1979. Once the railroad's hottest transcontinental freight, #200 in the last summer of operation is a sad collection of empty grain cars, empty copper concentrate hoppers, and a few containers and autos out of Tacoma. Only two SD40-2's up front and two cut in as helpers are needed to propel what's left of the Thunderhawk up Loop Creek Canyon./Photo by Blair Kooistra.

understood how pervasive the railroad's financial difficulties were.

In 1977, trains were running as they always had. Little had changed west of Harlowton except for dieselization.

Many of those who liked electrics lamented the change, and the passing of something that meant so much to them. Barry Kirk accepted the decision as an economic one, but sometimes he wondered if it could have gone the other way.

> "If the Arab Oil Embargo had occurred a year earlier, maybe the cost of the electrification would have looked better to them."

Kirk did note however, that the cost of electric power had unexpectedly risen along with oil, and that William Quinn, the Chairman of the Board may not like electrification. As Laurence Wylie looked back on the decision to de-electrify, he said:

> "The main problem was that the Milwaukee didn't have much money. In addition, the decision-making was in Chicago, a long way from the electrification. Most of the people at the top only came west a few times. They never really understood the electrification and they didn't like it. It broke my heart when they junked the electrification; that was my life."

ove:
veteran roundhouse rker in Tacoma's leflats facilities stares into space as he rem-sces about his 35 years h "The Company." 'l, 1978./Photo by ir Kooistra.

ht:
irence Wylie and the hor in March, 1977./ oto by Richard inheimer.

George Frazier, the Milwaukee's last Electrical Engineer, quietly presided over the removal of surplus equipment, and when that was done his department disappeared.

The end of 1977 hit like a bombshell. The Milwaukee netted a loss of $413 million and filed for voluntary reorganization under the railroad bankruptcy laws. The company was up to its ears in financial quicksand and still sinking. Such short term moves as sharply cutting maintenance and laying off employees were of no help. Faced with a desperate situation, the Trustee proposed a radical solution. Many branches and all track west of Miles City, Montana, would be abandoned. The 10,000 mile Chicago, Milwaukee, St. Paul and Pacific would shrink to a 3,900 mile Midwestern line, which was to be called the Milwaukee II. Many were surprised when this plan was not blocked by the Federal government or the courts.

The government, however, was worried. The Rock Island, another Midwestern carrier, had declared bankruptcy in 1975. An employee strike shut it down in 1979 and government action was required just to get trains running. President Carter directed neighboring carriers to provide service on the Rock Island, but the cost charged to the Federal government was estimated to be $3.5 million per week. If the Milwaukee's plan to cut back its route were blocked, the Company might declare itself financially unable to operate. In order to assure efficient movement of grain in the Midwest, the government could end up paying for operation on the Milwaukee as well as "The Rock." The government thus took a hands off approach.

The last train left Tacoma on March 15, 1980, and the 71 year old Pacific extension became history. Debates about the electrification became moot as the very tracks it ran on were pulled up or sold to other railroads. By 1983, most of the old electric line resembled a one lane dirt road and the employees were retired or scattered.

An era had ended, but another had begun. It was an era of the profitable giants in the railroad business. The strong became stronger as the Burlington Northern combined with the Frisco, the Norfolk and Western merged with the Southern, and the Union Pacific, Missouri Pacific and Western Pacific became one. The future held little for small, weak railroads and the Trustee of the Milwaukee II, the courts and the I.C.C. understood that. On February 19, 1985, the company was sold to the Soo Line, a subsidiary of the Canadian Pacific Railway.

Bitterroot Mountain right-of-way in the Adair Loop as seen on June 16, 1984. No track, no poles, just line-side sheds, a one lane dirt road and memories./Photo by Noel T. Holley.

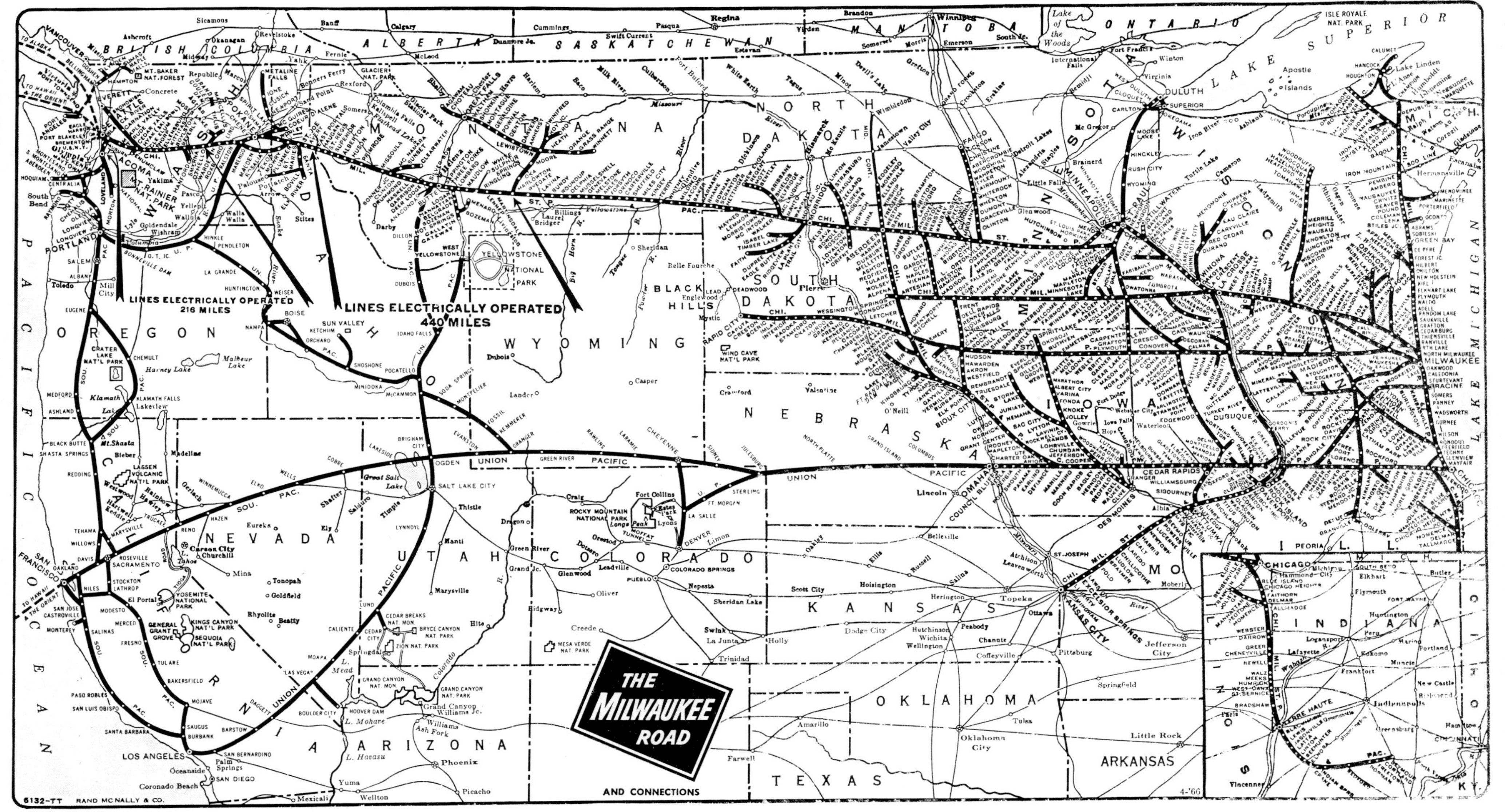

THE MILWAUKEE ROAD
AND CONNECTIONS
LINES ELECTRICALLY OPERATED 440 MILES
LINES ELECTRICALLY OPERATED 216 MILES
PACIFIC OCEAN
LAKE SUPERIOR
LAKE MICHIGAN
BRITISH COLUMBIA
ALBERTA
SASKATCHEWAN
MANITOBA
ONTARIO
WASHINGTON
OREGON
IDAHO
MONTANA
WYOMING
NORTH DAKOTA
SOUTH DAKOTA
NEBRASKA
MINNESOTA
IOWA
WISCONSIN
MICH.
ILLINOIS
INDIANA
OHIO
KY.
MO.
KANSAS
OKLAHOMA
ARKANSAS
TEXAS
COLORADO
UTAH
NEVADA
CALIFORNIA
ARIZONA
SEATTLE
TACOMA
SPOKANE
PORTLAND
SAN FRANCISCO
LOS ANGELES
MINNEAPOLIS
ST. PAUL
DULUTH
SUPERIOR
MILWAUKEE
CHICAGO
OMAHA
KANSAS CITY
DES MOINES
CEDAR RAPIDS
DENVER
SALT LAKE CITY
OGDEN
UNION PACIFIC
YELLOWSTONE NATIONAL PARK
BLACK HILLS
6132-TT
RAND McNALLY & CO.
4-'66

Soo Line Railroad Company

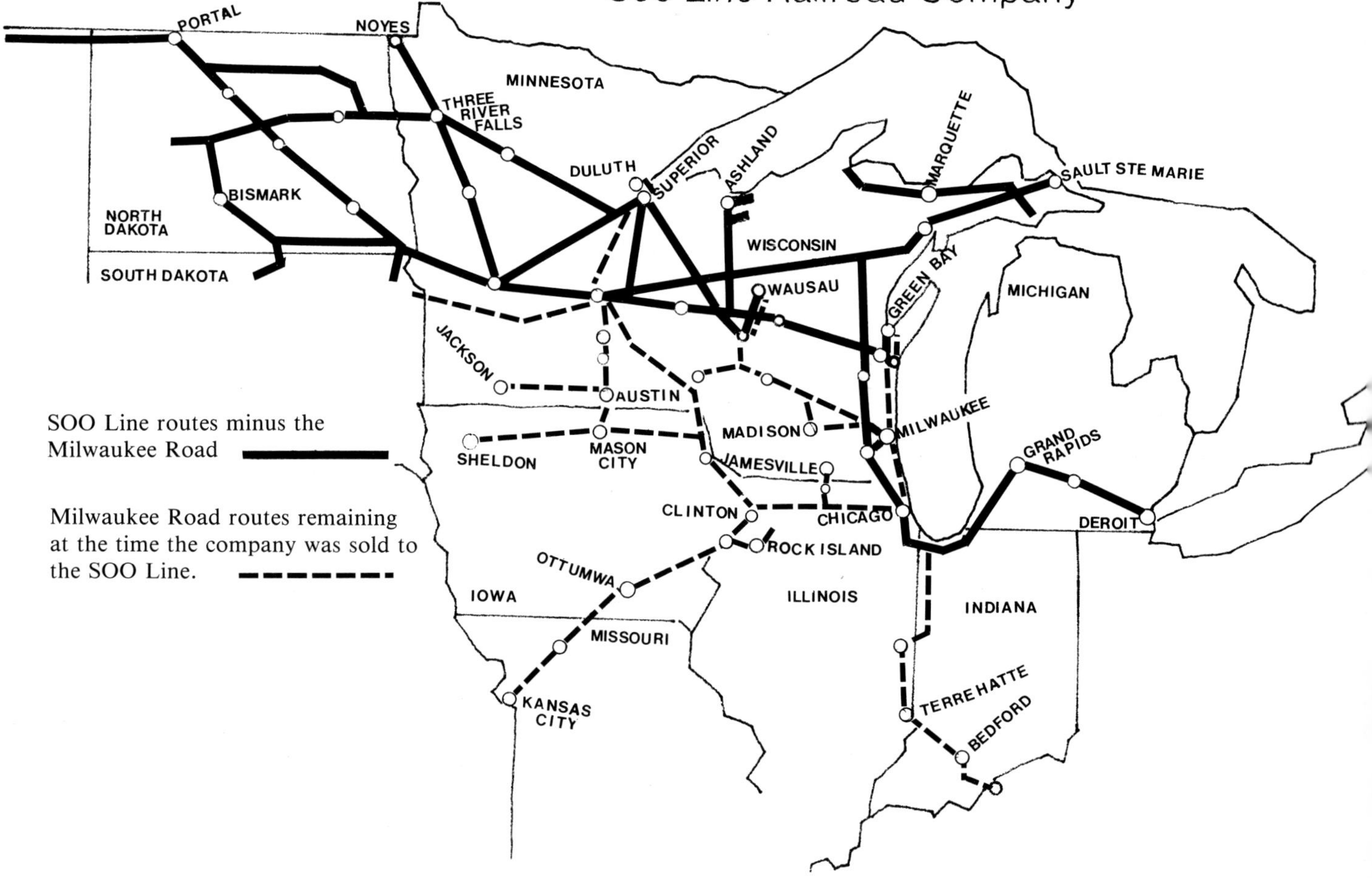

PORTAL
NOYES
MINNESOTA
THREE RIVER FALLS
DULUTH
SUPERIOR
ASHLAND
MARQUETTE
SAULT STE MARIE
BISMARK
NORTH DAKOTA
SOUTH DAKOTA
WISCONSIN
WAUSAU
GREEN BAY
MICHIGAN
JACKSON
AUSTIN
MADISON
MILWAUKEE
GRAND RAPIDS
SHELDON
MASON CITY
JAMESVILLE
CLINTON
CHICAGO
DEROIT
ROCK ISLAND
OTTUMWA
IOWA
ILLINOIS
INDIANA
MISSOURI
KANSAS CITY
TERRE HATTE
BEDFORD

SOO Line routes minus the Milwaukee Road ▬▬▬▬

Milwaukee Road routes remaining at the time the company was sold to the SOO Line. ‒ ‒ ‒ ‒ ‒

From 1910 until 1980, the Milwaukee Road operated in excess of 10,000 route miles of trackage. The map on the preceding page shows that route. In addition, it emphasizes the Milwaukee's connections with the Union Pacific and the Southern Pacific. Between 1980 and 1985, large portions of the system were sold or abandoned in an attempt to preserve the railroad's profitable core and avoid total bankruptcy. In 1985, when the Milwaukee was sold to the Soo Line, only a shrunken route structure remained. Much Midwestern trackage was gone and the old Pacific Extension had been cut back to Ortonville.

TALES OF THE RAILS

Tales of the Rails is a collection of railroad memories. While many concern events which were far from ordinary, they serve in their own way to illustrate what life was like on the western end of the Milwaukee Road. My own story and one by Don Dietrich capture what was seen and felt by non-railroaders to whom a cab ride was an entirely new and different experience. The balance of the chapter is stories from railroad employees. They are the stories which railroaders tell and laugh about when they recall the good old days. They concern events seldom seen by outsiders and never retold by the corporate public relations office. All of the stories are true and since the employees mentioned have long retired, it was not necessary to conceal identities.

Train #264 between Kyle and Stetson on May 15, 1974. E73 and E72 lead diesels 4005 and 4016 on 3830 tons of freight./Photo by Ted Benson.

E78 WEST, AN ELECTRIC JOURNEY

By Noel Holley

In the days when the electrics still ran, I was fortunate enough to get a ride in the cab of a Little Joe. Soon after the trip, I wrote about what I had experienced, so that the passing years would not blur the details of what I saw and felt.

It was March 1973, a winter when the customary western snow packs never came, the days of electric operation were numbered, and the line was in terrible shape. The Milwaukee's luck was coming in the fabled threes, and in the last three days there had been three derailments on the same stretch of line.

Two were up on mountainsides where the lines snake through the Bitterroots, and one was down along the St. Regis River.

I had stopped in Alberton to see Adam Gratz, an old friend who was an engineer. He offered me a ride and I quickly accepted, but it was hope against hope that our train would have an electric on it. The wrecks had ripped out the trolley in two spots and few electric locomotives were being sent west of Deer Lodge until it was repaired.

The usually empty Alberton yard was full as three west bound trains waited for clearance to proceed; only one had an electric on it. After a few hours of waiting, word came that we were second out and our train was E78 west. We had a Little Joe and three diesels to roll our 70 cars of freight from Alberton up across the mountains to Avery.

In Alberton, the depot sat at one end of the yard. All three trains were yarded with the engines coupled behind the cabooses. This way the engine crews didn't have to walk a mile to get to their cab, they just came out of the depot, climbed into the engines and ran them around to the head of the trains. This isn't standard practice, but then trains didn't usually stop and wait at Alberton. They usually just made a five minute stop at the depot for an engine crew change and then rolled out at 5 mph until the rear crew had changed.

About 3:00 p.m. a call came saying the line was clear and the first train, extra-3014, began to roll. We climbed into E78 and raised the pan. The brakeman uncoupled us from the caboose and we backed the engines out onto the yard lead and then ran slowly around the train. We coupled on ahead of some high cars and tested the air brakes. They were working fine, so when the signal up ahead showed green we began to roll. The cab rocked slowly like a boat, and from the tall and armless third chair I saw a panorama. There were pastoral farms with winter-brown grass, and evergreen forests on the mountains.

The three diesels behind us thundered with a rasping chant, but the electric we rode in was silent. The diesels alone were enough to haul the train on this trip and the Joe was basically a helper. Though the pan was up and we could use the Joe if we needed to, we didn't need it yet. Adam wouldn't be working the Joe until after we passed the wreck at Henderson. The throttle arm of the Little Joe's controller rested in the "off" position as Adam accelerated our train. He notched out only the miniature diesel throttle which was mounted beside that of the Joe. The lever which connected the two throttles for simultaneous control lay unlatched and unused.

As we rolled along the Clark Fork Valley, Adam called out the block signals ahead. "Green board," he said and Bill Reeves the brakeman called back "green board," in agreement. Our path was like a steel boulevard. The Milwaukee spaced its trolley poles 150 feet apart and to ride along them in a locomotive was like riding down a tree lined street. Some were warped, all were stubbed, and by their color they showed what 50 years of heat and snow can do to wood.

"Henderson," (Montana) said a railroad signboard. Our Little Joe electric glided through at 5 mph with the pantographs down and the geeps behind us working.

Twisted freight cars lay on both sides of the track. The trolley hung in crazy swooping sags and then disappeared in the trees on the mountainside. Men in white hardhats were everywhere and the trolley crew was already setting new poles to replace the broken ones.

A few days before, a rail had turned over beneath an eastbound freight and tied up the main. Wrecking crews with dozers and big hooks had plowed out the right-of-way and now we were rolling through.

A few mileposts down the track our slow order expired and we passed an air-gap in the trolley. Adam pulled a switch and the forward pantograph gracefully rose up into its diamond shape. It did a springy, slow-motion bounce and then slid along the wire. Another switch was thrown and the blowers started. Adam hauled back on the throttle and the train picked up speed rapidly. Soon we were charging along at 50 mph to the roar of the blowers and the clanking of assorted metal parts.

The train was thundering along toward Haugan and a meet. As the cab rocked, I held on tightly to my seat and wished that it had arms. The cab seats were up on pedestals and a long way from the floor.

Adam slowed the train as we appraoched Haugan and we rolled into an empty siding. As we rolled past the tiny Haugan depot, the agent hooped up some orders. Bill leaned out his window, grabbing them on the fly and we kept on rolling until our entire train was clear of the mainline.

A rainy May 14, 1974 portends thicker precipitation on St. Paul Pass as #264 gets ready to go at Avery behind Joes E78/E21 and SD45 4009—not much margin for failure with 96 cars, 4431 tons eastbound. Joes have all four pantographs raised to cut ice on the overhead above Avery./Photo by Ted Benson.

View from the cab of E78 while exiting a tunnel west of Alberton./Photo by Noel T. Holley.

As we waited, Adam and Bill talked and shuffled papers. They matched and rechecked our orders and then settled back into their chairs. There was nothing to do but watch and wait. The blowers roared and the diesels drummed, but they served only to mark time in a seemingly unending motionlessness.

Something caught my eye. A bright light in the distance. It held steady and then, as I watched, it grew larger. Behind it loomed a string of geeps raising dust. First, a sound like bass drums and then the swaying shapes, the rush of the cars and the whining wheels rhythmically clacking at the rail joints. It passed and receded into the distance, the quiet returning as the clacking left, and the block signal changed from red to green.

''Green board,'' agreed the train crew and we were off. At Haugan the grade began, and I could hear the engines working harder as we climbed. The Little Joe growled like a giant streetcar and the droning diesels thundered. We snaked along shadowy mountain ledges and crossed through cuts and over trestles. At every curve the flanges squealed and the beautiful scenery was everywhere. The sun was getting low and the dim light made photographs from a moving train impossible.

It was snowing now and the snow that was landing contrasted sharply with the black splotches in the roadbed beside the flange oilers.

As our train got higher in the mountains, I began looking over the cab. The E78 had been wrecked once and the cabs had been rebuilt with diesel parts. There were few alterations. Fewer cab side windows and chrome EMD grill panels on the air intakes were the most noticeable differences.

According to the voltmeter in the cab, we were only getting 2500 volts as we climbed the eastern side of the Bitterroots. This didn't slow our progress any though. We were probably making a good 35 or 40 mph on the 1.7 percent grade.

"E78," crackled the radio. "East Portal to E78 west."

"What do you need," responded Adam as he turned on his mike.

"There is no clearance given for the trolley on the west side of the hill. Be prepared to drop your pantograph at the East Portal air-gap."

"We need the Joe for braking," Adam groused.

"See if you can get clearance."

The East Portal operator responded that he would try. We rolled onward in suspense and waited for another call on the radio. When it didn't come, Adam called East Portal back.

"Have you gotten clearance yet?" he asked.

"Nope, I can't raise the trolley crew on the radio. I'll keep trying though."

"If you reach them, ask if they will sectionalize the part of the trolley they are actually working on and let us use the rest."

"I'll see what I can do," came the response, and with that East Portal clicked off.

The trolley was designed so that it was possible to turn off stretches of trolley without affecting other areas of the line. To make this possible, one stretch of trolley was insulated from the next by air-gaps that the electricity cannot bridge. These air-gaps were placed regularly along the line so that there was no need to shut down the whole railroad in order to repair one stretch of wire.

It was almost dark when we rounded the last curve before East Portal. Looking up the track, we could see the huge red brick substation building and the row of houses for the men who lived at that isolated spot.

A few hundred feet before the air-gap stood the East Portal operator with a yellow flag. We dropped our pantograph and glided slowly up to him. He hooped up some orders and waved. We waved back and rolled beneath the dead trolley wires and into the tunnel at St. Paul Pass. The east portal of the tunnel lay at the convergence of two mountain slopes. A hundred foot snow shed stretched outward from the tunnel to protect the entrance from snow slides. Inside, the tunnel still showed the form marks from when its concrete was poured. Wooden sheets had formed the sides and heavy planks the crown. The tunnel was hewn out of rock, and where the rock was firm there was no lining. Water incessantly dripped and poured from the ceiling. It ran down the walls and streaked the soot that had been left by the passage of the steam engines and diesels that roamed the line. The catenary hung from "U" shaped brackets on the ceiling, and our headlight illuminated the two parallel contact wires as they stretched off into the dark.

When we finally left the mile long tunnel, darkness had come and we now had only our headlight to help us see down the winding 1.7 percent grade to Avery. It was snowing and the big electric was silent. The diesels were set in dynamic braking and the loudness of their chant was greatly diminished. They reminded me of a truck going downhill on compression as we squealed through the tight mountain curves.

E78 on train #264 meets train #261 headed by SD40-2 3025 at St. Regis on May 14, 1974./Photo by Ted Benson.

Usually the Little Joe would be helping to slow the train by using its 5,000 hp in regenerative braking; however, doing this would pump several thousand kilowatts of direct current back into the wires. Tonight the trolley crew was repairing damaged wires at the wreck and they didn't want the hazard of working on them live.

Our orders stated that we would have to stop and wait uphill from the wreck behind extra-3014. We would be contacted by radio to tell us when to proceed.

Several times Adam tried to contact the trolley crew, but all to no avail. The only answer was static.

As our train sat and waited on an earthen fill, the cab got colder. The cab heater was electric, but we weren't able to use it. We had to run our four beam headlight and the locomotive control circuits on the batteries. The batteries might run down too fast if we used the heater too. There was a small coal stove in the cab, but there was no coal so we couldn't use that either.

After about an hour, the trolley crew crackled onto the air and told extra-3014 to proceed on down the hill. A few minutes later we got an OK to proceed and we were on our way. We glided past a block signal and no light showed.

"Unnecessary," grumbled Adam. "They could run the signals above the wreck from East Portal and those below it from Avery. There is no need to shut off all the power to the whole hill just to fix one small section of the line."

E21 has left her train at the Avery depot. She is headed for the engine terminal./Photo by Ted Benson.

"E78 to trolley crew, E78 to trolley crew," he called on the radio. "Trolley crew," they crackled in response.

"How about sectionalizing the wires you are working on and then letting East Portal and Avery put the rest of them back on the line?"

"Can't do it," came the response and there was no explanation offered.

We rolled on in silence with the yellowing headlight to show us the way.

After a while of uneventful riding in our cold and heatless cab, Adam craned and stared down the tracks.

"There you can see where they derailed," he said and pointed out the window.

I anxiously stared ahead, but saw nothing. There was darkness and snow, but no catastrophic signs that a train had left the rails. Adam and Bill looked ahead intently, but I was at a loss to know what they were staring at.

When I asked they pointed, and this time I noticed. In each tie there were dents and gouges between the rails. They were not deep, but they were quite visible and they stretched off into the murky gloom ahead of us. Obviously, a freight car deep in the train had jumped the track without leaving the train or coming uncoupled. I watched and waited like someone following the trail of a wounded animal. The dents and gouges continued for what seemed to be miles and then we rounded a curve at Kyle.

Here the track branched into two sidings in addition to the main and we crept through on the rebuilt main. Where the two sidings had been was now just steel spaghetti, wooden tie chunks and freshly tilled earth; container cars askew on the high side and a path where something had slid across our ledge and plunged down the slope. Farther on were two boxcars standing alone.

There were newly set poles, and the trolley darted from above track center to near the poles on the curve's inside. The trolley crew was fixing it slack and they would draw it tight when they pulled it back into position.

Searchlights were set at strategic places, their beams illuminating the rails, wires and men in the snowy, tree filled darkness. There was a huge roaring bonfire of tie chunks and around it huddled most of the track gang and trolley crew. They were out of the way of our passing train, but I shivered to think of their work. This was their third wreck and they were working around the clock to keep the line open. Cold mountain days and colder winter nights were their constant enemies as they worked against the clock to repair the line.

While rolling along the main, we passed a long cut of cars which was standing on the nearest siding. As we did the gloom returned and the lights and fire fell behind us.

After passing the wreck, time went fast. Our speed climbed back to somewhere near 30 and the train seemed to drop quickly into the wooden canyon. Soon we were nearly at creek level and slowing to the chant of the diesels and the clanking of assorted metal parts. We rounded the bend where Motor Creek joins the St. Joe River, and the Avery yard fanned out ahead of us.

The lights of an eastbound freight shone out through the rain as it waited at the depot. We passed a trolley air-gap and our pantograph rose back into a diamond shape beneath the wires.

As we drew up to the depot, Bill said goodbye and climbed down the ladder to the ground. He threw a switch, raised the pin lifter to uncouple the Joe, and walked off through the rain to the depot. The depot looked warm through the rain and the cold, and the substation beside it sat with a kind of church-like dignity, a rainwashed cathedral from which rotating converters and transformers filled the night with their sounds.

Adam opened the throttle and our big electric locomotive glided effortlessly toward the engine terminal. The diesels and the rest of the train were left where they stood. Another train crew would soon be out to take them to the coast.

Our Little Joe ground to a stop beneath the sand tower and both Adam and I climbed down. An adventure ended, but I'll always have its memory.

SNOW IN THE PASSING TRACK
As told by Adam Gratz

One winter the weather kind of moderated while we still had a lot of snow in the mountains. The warm weather brought a lot of snowslides and they were using a rotary to keep the line open. I was making a run to Avery one day that winter when they ordered me into the passing track at Stetson. They were trying to get some trains out of Avery behind a rotary that had just gone by.

Stetson was a passing track about five miles out of Avery that they actually didn't use very much. Whenever they would plow out the mainline they would plow it with a cut widener and a flanger and they would plow the snow right into the passing track. They hadn't plowed Stetson all winter and so it was piled full of old packed snow.

When the dispatcher ordered me to head in there, I argued against it. I told him that with all that turned over snow in there I didn't think I could get in. I might end up just sitting there tying up the mainline with my train. Even after I told him that, he didn't believe I would have any trouble so he ordered me to head in anyway.

I had a double Joe and a small train of 73 or 74 cars so my train was just short enough to fit. I struggled into the clear and made the meet, but when I got ready to leave, my train wouldn't move. I was stuck. I gave her full throttle, put her up to about 400 amps but I couldn't move anything. I knew if I could free the engines I could get the train out, so my brakeman Milt Rancourt went back to uncouple us. With the train stretched out tight he couldn't uncouple anything, so I decided to back up and get some slack. Well I put the engines up to 400 amps in reverse and nothing moved then either. He was standing back there waiting, but I couldn't even get the pin on the first car. While I was sitting there waiting for a meet with that train, the packed snow underneath the warm traction motors must have frozen up. It froze the engines tight to the rail. I kept jockeying around with the engines, putting them in forward and reverse, trying to get out and after about an hour they finally broke loose. I guess it was the heat from my traction motors and the air from my blower vents that finally thawed the ice underneath the Joes. Anyway, I got my pin, ran out and cleared a path through to the mainline, and finally got my train into Avery.

An eastbound freight has arrived from Othello and E75 has coupled on for the trip to Harlowton./Photo by Gordon Rogers.

Left to right: Engineer Adam Gratz, Fireman Dan O'Bannon, Traveling Engineer Linton Shaw./Photo from Adam Gratz Collection.

DRAWBARS AND GARBAGE CANS
As told by Adam Gratz

At Avery one time we had a lot of snow, about three feet on the ground or maybe a little better. The County had been plowing the streets and they would pile the snow up on the outside of the road next to the mainline. Anyway the streets were getting kind of narrow because of the snow piles up around them, so Tom Prata the Road Master thought he would do the County a favor. He told them that if they wanted to plow the streets real wide, they could plow the snow onto the tracks. Then we would throw it over into the river with the rotary snow plow.

I was working the rotary the day they widened the street. We had just cleaned out the west yard, so we came up to the east end and started to plow that yard too. We started with the mainline so we could work toward the river and we done real good until we got down by the tool shed. All of the sudden that old rotary made a big clank, jumped about a foot in the air, and scared the daylights out of all of us. We stopped to investigate and found a big drawbar, including the knuckle, lying on the track. There used to be a scrap pile next to the tool shed but Bernie Benson, the grader operator, didn't know that so he plowed it all onto the mainline with the snow. We dragged that drawbar off of the track and started plowing again, hoping that there weren't any more drawbars under the snow. There weren't any more but, when we got down to where the apartments we called the flats started, we heard the darndest racket you ever heard in your life. You would see a big dark streak going out through the chute toward the river. Pretty soon a beat up garbage can would come pouring out of there and then the next thing you know there would be another garbage can. It was like that all the way up to the substation. All the people in the flats used to set their garbage cans across the road next to the mainline and I guess Bernie forgot the cans were there. He widened the road all right, but in the process he pushed all those buried garbage cans onto the mainline.

The people living around those flats had to rustle new garbage cans for the rest of the winter. There wasn't much left of the ones they had.

Steam rotary pushed by a Freight Motor. Date unknown, location may be Saltese, Montana./Photo by W.H. Edwards.

THE RUNAWAY
As told by Swede Hansen

In the 1950's, I was firing in passenger service between Deer Lodge and Avery. One night in January of 1951 I made a westbound run with an old engineer named Casey Hyde. We were westbound with the E21 on Number 15, the Olympian Hiawatha. The timetable had us out of Deer Lodge at 6:55 p.m. and we were due in Avery at 11:30.

The trip was going real good that night, but it was snowing and raining all the way. The snow was heavy and exceptionally wet, so it was sliding down wherever we had steep cliffs or cuts. West of Haugen we had a little slide here and a little slide there, but it was nothing really to worry about.

Part of a fireman's job was to patrol the engine and inspect all of the equipment about once every half an hour. Near 11:00, when we got to Adair, I was due to make my last patrol, so I started on my rounds back through the engine. I was in the back end of the Joe checking the Elesco train heat boiler when all of a sudden we hit a slide. I felt the engine jolt and jerk and I thought we were headed down over the bank. I didn't want to get burned so I jumped into the back nose to get as far away from the boiler as possible. I opened the back door and looked out, and just as I did, the Joe cut off from the train. Just what pulled the pin lifter I didn't know. It may have been a rock or a tree, but there was the train stuck in the slide and we were still rolling. The slide must have come down from up high, because it was full of rocks and little Jack Pines. As the Joe went through the slide, one of those Jack Pines swiped the pantograph off the roof. We lost all the air too as the brake equipment was torn off one side of the engine. We were left with no power and no brakes.

I don't know why I didn't jump off the back end of the Joe when it left the train, but I just didn't. I thought that Hyde might be up in the cab hurt so I didn't have time to get scared. I just took off running through the passageway to the front of the engine. I got banged around pretty good in there and I cut my head on something, but I didn't stop to worry about it. When I got to the cab I could see Hyde wasn't hurt. He was just standing out in the nose of the Joe trying to figure out how to work the hand brake.

Hyde was an "old head" up in his seventies and a real nice fellow to work for, but he relied on the fireman for so damn much that he had never set the hand brake on a Joe and he just didn't know how. He was standing there looking at it while we picked up speed.

I looked out the door just about then and when I did, I saw sparks coming up from the sides of the rails towards the front of the Joe. That is when I realized the pilot truck was derailed. It was just bumping along the ties, but it didn't slow us down any because all of the other wheels were still on the track. As we came up on Clear Creek bridge we were going 45 or 50 miles an hour and I thought we were going to leave the track for sure. When we got out on the bridge and I saw we would get across it, I went down into the nose and started turning the brake wheel. With that derailed pilot truck, we knew we wouldn't make it if we hit the east switch at Falcon. Since we weren't far from Falcon, I had to work fast.

It seemed like quite awhile before anything happened, and we were inside of tunnel 27 before I heard the brakes start to squeal. I just kept pulling on the brake wheel and tightening up the pressure after that. I tore the ligaments in my back from pulling so hard, but I was scared and I didn't know I was tearing ligaments. The brakes kept squealing and finally the engine stopped. We were inside tunnel 28 when I stopped it and Falcon was less than half a mile away.

E20 and a team of diesels haul an eastbound freight across one of the many Bitterroot Mountain bridges in the Adair Loop./Photo by Noel T. Holley.

After we had been there for quite awhile, the Conductor, Jack Webb, showed up. It took him quite awhile to get to us because the snow was wet and knee-deep and we were nearly two miles away from the train. He figured he would find us piled up somewhere, so he had started walking to the nearest telephone. Charlie Rock, the head brakeman showed up too. He was back in the train when we hit the slide. He said that when the train started jerking he stuck his head out to see what was going on. That's when he saw the Joe leave the train.

Luckily, the train stayed in the slide and the passengers were safe but we all must have sat there for seven or eight hours waiting for someone to come get us. They brought a crew and a freight motor up out of Avery to bring the E21 and the train down.

I was the only one injured in the runaway. I was out of service for pretty near a month with my back, but I had a hell of a time getting the railroad to pay me for my time off. They didn't want to pay me, but they decided they would after awhile. Then they decided to give me a gold pass in recognition of what I did in stopping the engine. With the gold pass I could ride the passenger trains for free. It wasn't worth nothing but they thought it was at the time.

I'm 71 now and I don't remember everything about the railroad, but I remember the runaway. It scared the devil out of me. I was so scared I actually prayed in the runaway E21. I thought I was taking my last ride on the railroad.

Young Deer Lodge Hostlers Stan Christenson on the left and Swede Hansen on the right./Photo from Swede Hansen Collection.

A HOPPER LOAD OF TROUBLE
As told by Bill Lintz

On the night of November 9, 1957, I was running a couple of Joes and a westbound freight from Deer Lodge to Alberton. Everything went fine for the first half of our trip, but about one mile west of Iris I got a call on the radio. Rex Logan, the Conductor, wanted to know if I had the brakes set on the train. When I told him "no," he said we better stop and look for dragging equipment. "There's sparks flying out of the back of this train."

I stopped about 2 miles west of Iris across from the Rock Creek Tavern and we all got out to inspect the train. Since we couldn't find anything we decided to proceed, but the sparks kept flying. When we passed the station at Missoula, the operator there saw all the sparking and flashing so he came out to take a look. He found some broken rails in front of the station but it was awhile before he could do anything about them, by that time we were in Alberton.

Rex stopped us several more times to inspect the train for trouble but we never found anything wrong. We got into Alberton at four or five o'clock in the morning, and that is where the train changed crews. We were standing near the station when the train went by us and in the middle of the train here came a big hundred ton coal hopper, kerTHUMP, kerTHUMP, kerTHUMP. It had a chunk broken out of a wheel and that left a big gap in the wheel tread. It was a heavy car and every time the wheel came around it would break the rail on one side. We had broken over fifty miles of rail with that car. The rails carried the return current from the electric engine to the substation and the sparks Rex Logan saw were caused by the electricity jumping those gaps in the rail. We hadn't found the broken wheel because we hadn't made a good enough inspection. All we could think of was something dragging and we didn't find anything.

Shortly after we got to Alberton, #16, the eastbound early morning passenger train was due. By this time we realized we might have damaged things with that flat wheel, so we called the dispatcher and told him. He gave #16 orders to proceed at restricted speed and to be prepared to stop short of any obstruction. #16 only got as far as Soudan, about five miles out of Alberton when the third passenger car behind the Joe caught a wheel flange on the broken rail. The rail went over and the cars just bumped along the ties tearing up the track. No one was injured because they were going so slow, but the wreck tied up the line for a couple of days. They had to send out the wrecker to clean that mess up.

Work trains were sent out to put angle irons on all of the breaks in the rail. It took them quite awhile because there were so many breaks. They were spaced at the circumference of a wheel all the way from Iris to Alberton. It took them months to re-lay all that rail. The work ran into December and January, when the weather turned bad, so they had to work in snowfall and ice. Eighteen inches of snow fell after the work trains distributed the spikes, bolts, washers, tie plates and everything they needed to tie down the rails. The track gangs had a hard time finding all those parts buried under the snow so the railroad had to bring in a lot of extra men to brush and sweep the snow away. They found the materials but it took them quite awhile. In the meantime there were slow orders on the track. You were limited to 10 mph for 55 miles.

It was a mess, but the railroad didn't lay the blame on us. We had taken every precaution we could under the circumstances.

Bill Lintz running the Northern Pacific Golden Spike Centennial train at Deer Lodge in 1983. The train ran on abandoned Milwaukee rails because the Burlington Northern would not participate in the celebration or allow the train on their tracks./ Photo from Bill Lintz Collection.

*Cleaning up a wrec Soudan, Montana. unknown./*Photo Barry Kirk.

LIGHTNING BOLTS AND ARC FLASHES
As told by T. Barry Kirk
Retired Electrical Engineer

One stormy night during the 1930's, lightning struck our high voltage transmission line near Two Dot, Montana. Lightning packs millions of volts of electricity and is very unpredictable. It went down the transmission line, into the Two Dot substation and caused the transformers to explode. The explosion blew out the roof and part of the back of the building. The transformers were oil cooled, and the oil that was in them burned for several days. Rainier Beeuwkes, who was head of the Electrification came out by train to look over the damage, and the building was still burning when he got there. He was anxious to know what the extent of the damage was so he told me to climb up a ladder to the roof and look down through the hole for him. With smoke still pouring out of there, I wasn't about to go up on the roof. I told him that if he wanted to know what was going on in there, maybe he should go up on the roof and look for himself. He decided to wait until the fire went out so he could look through the windows.

When there was frost on the trolley, our motors would make huge arc flashes as they went down the line. One time during World War II that caused us some problems with the government. After Pearl Harbor was attacked there were blackouts every night. The government didn't want to see any light coming out of houses, the street lights were turned off and there were headlight hoods on the locomotives to keep the light from shining up into the sky. One night when there was a lot of frost, a Milwaukee freight lit up the whole countryside from Tacoma to the crest of the Cascades. The government people were very upset about that. They said they didn't want us to run during blackouts if we couldn't keep the arcing under control.

HOLES IN THE ROUNDHOUSE WALL
By Wade Stevenson
Retired Roundhouse Carpenter's Helper

Many times the Bipolars and Freight Motors ran through the back wall in the Othello shop. They rolled very easily and sometimes the crews would drop them in on the fly without having enough air in the brakes to stop them.

One night after #18, the Columbian, came into Othello, they took the Bipolar over to the roundhouse to turn it. A mallet spotted it on the turntable and then gave it a nudge so the hostler could get the pin. When they nudged it the Bipolar went sailing off the turntable and through the roundhouse. There was no way to stop it because there was no air on the Bipolar and the hostler didn't have the handbrake set. The pans were cleaned off as it knocked out the sills over #7 door and broke out the overhead window glasses. Then it tore down the smoke jack and out it went through the wall of the shop, ending up against a dirt bank. When I came on duty that night, the hostler was already out of a job and drunk.

The electrics were designed so that you could bring them into the shop by plugging a long cable into them and operating a switch at an electrical box. Men dropped them in on the fly because it was always such a job dragging that heavy cable around.

Wade Stevenson at Othello, Washington, 1953./ Photo by Richard Steinheimer.

MEMORIES OF THE STEAM DIVISION
As Told by Paul Edmunds

In 1943, I went to work for the Milwaukee as a fireman. I fired for four years before I made engineer, and during that time I fired for some interesting guys. Cortez Burrell and Eddie Maxwell were a couple of them.

Cortez Burrell was an old boomer who scared the daylights out of a lot of people on the Idaho Division. He was an engineer who really liked to run fast, and when some of the younger fellows came up into the cab of his steam engine, he would ask them if their life insurance was paid up. One time somebody asked him how he knew whether or not he was going too fast, and his response was, "If I look over and I can't see daylight under the seat of my fireman's pants, then I'm not going too fast."

Out on the road, Cortez liked to dog along coming up Plummer Hill and through the hog backs to Worley. After the stop at Worley, he would take off down through Mozart like the devil was after him. One of his favorite tricks was to yank the throttle open and then get up and go to the gangway to take a leak. He would stand over there quite awhile and some firemen would get nervous waiting for him to come back to the throttle. I was firing for him once when he did that. We got going so fast down a sag that I went over and set the airbrakes. He came back in the cab about then and said with a laugh, "What's the matter kid, we going too fast for you?"

Another guy who liked to run fast was Eddie Maxwell. If he was running passengers and his train was delayed 15 to 20 minutes, he would get very antsy. Sometimes in Othello we were delayed while they loaded express and mail from the depot. Maxwell's voice would get high pitched and squeaky as he kept asking me "What are they doing now." I would look out my side of the cab and tell him they were still loading. Then he would come look for himself, and all the while telling me that we were late. When he finally got the highball you had to be ready to go with the engine "hot and full of water." He would really give the engine a mauling. All the way to Warden. Then it was down a gentle grade to Roxboro and across the flat to Servia. He was at it again until you got to Hillcrest. It was down grade to Marengo, where we took water and got onto the Union Pacific tracks. Maxwell wasn't happy until just west of Cheney where he could look across the cab and tell you, "On time at Geib." He had such a reputation for speed that the Union Pacific was on the lookout for him whenever he came onto their track. Maxwell was good, in terms of speed; he could get everything out of the track that was possible, and he wasn't reckless. He was always careful and under control. The speed limit on Mica Hill was 30 mph because of the curves, and Maxwell had no trouble coming downhill above the limit. He would set some air on the train brakes and work a heavy throttle against the air. His train would come down at a steady speed, stretched out like a snake and trailing sparks.

267 a Class S-3, 4-8-4 is eastbound with the Columbian near Dishman, Washington, October 2, 1952./Photo by R.V. Nixon.

You can run a lot faster with the train stretched out than you can with the slack running in and out.

The U.P. knew from the crew registers who was coming into Spokane, and anytime Maxwell had that run, they were camped out waiting for him. I was firing for him one time when the U.P. decided to make sure he slowed down. We were on a westbound run to Spokane and entered U.P. territory at Manito, the summit of Mica Hill. We were moving pretty good down near Freeman when we heard a couple of torpedoes go off under the engine. The rules said you were supposed to slow down to 20 mph for one mile when you heard them pop. They were used to alert you about trouble ahead on the line and you had to be prepared to stop. We knew there was no trouble, the U.P. just liked to know you were paying attention and they liked to see sparks fly as you put on the brakes. The torpedoes were a warning to Maxwell about speed. When they went off, I looked out the cab window and saw a car on the highway following us. We knew it must be a Roadmaster or one of their other officials checking up on us so Maxwell slowed down. The U.P. was strict about enforcing the rules and they would write up Milwaukee men as well as their own. All the while we were beside the highway, Maxwell kept asking me to check and see if that car was still out there following us. The highway ran beside the track for 12 miles all the way to Dishman, and he couldn't speed up if the officials were watching him. When we got to Chester the car turned off. Chester was the bottom of the hill and the speed limit was 50 mph from there on into Spokane. Maxwell rolled the train into Spokane within the speed limits that night. Even after we lost sight of the highway there might have been people watching us. He stayed true to his reputation though. You would still find him speeding anytime he didn't think he was going to get caught.

Snow wasn't a big problem on the Idaho Division, but sometimes in the winter we used to have a lot of trouble with snow drifts around Manito and Worley. The winter of 1949-50 was a rough one. It snowed a lot that winter. Bucking snow on a steam engine was miserable. When you hit a drift, the snow would pack all over the front of the engine and come over the top. If the roof vent was open, it would be like a turkish bath in the cab as the snow dropped down on the backhead. You had to keep the vent and the windows shut to keep the snow out. If you were going to hit a drift, you had to close the front damper and open the rear one. Otherwise, snow would fill up the firebox and put out the fire.

Locomotive #250 or 251. Date and location unknown./Photo by Laurence Wylie.

PEAK LOAD ON THE MOTOR DIVISION
As told by Pat Chester

I hired on as a fireman with the Milwaukee in 1943. I got a job firing steam engines on the Idaho Division for $7.40 per hundred miles. That was a day's wages then and it was a lot more money than I was making on my previous days' job.

I was just a kid of seventeen when I started out and I barely weighed enough to pull down water spouts when the engine stopped for water. Sometimes the engineer would have to come up on the tender to help me. I had experience running boilers in a dairy before I went to work for the railroad, so it didn't take me long to learn to keep the steam up on a locomotive.

I stayed on the Idaho Division until 1954. After the diesels came, they didn't need as many crews on that division, so I ended up being cut off half of the time. I had a family to support and I couldn't do it if I wasn't working, so I transferred to the Coast Division. I could work regularly on the "Motor Division."

Most steam engine men who didn't work on the Motor Division didn't want any part of electrics. The ones who did work with them learned to like them. I didn't know anything about them when I took the transfer, but I wanted to work and I was willing to learn. When I left Idaho, the old timers wished me good luck and reminded me that before the Coast became electric, it was a steam division. The first guys on it used their knowledge of running trains with steam to help them run electrics. I could do that too, said the old timers.

Hostlers Pat Chester and Bob Donley at Othello, Washington around 1950./Photo from Pat Chester Collection.

On an electric, a fireman was an apprentice engineer and had responsibility for inspecting and maintaining all of the equipment on the locomotive. I had been firing on the Coast Division for about a year and learning what I could when an engineer's job came open. Because of the time I had in on the Idaho Division I had enough seniority to bid on it and get it. One afternoon, a couple of years after I was promoted, I was getting ready to make a westbound run from Othello to Cle Elum. Just about then my conductor, Melvin Faudren, came up and asked me if I would forget about the rules on stopping for Peak Load Time. He wanted me to make a fast run. He was mayor of Othello and he had a speech to give that night. He wanted me to get into Cle Elum early so that he could catch a taxi and get back to Othello in time for his speech.

I was just a young guy then and rules on Peak Load Time didn't mean much to me. They were just more rules made by people who didn't run trains, people who had no sympathy for guys out on the road who might have to stop trains on the hill to abide by the rules. Those guys would make rules and we always tried to figure out how to get around them. Peak Load Time was between about 5 pm and 6 pm in the winter. It was the time when Washington Water Power had its biggest load. They didn't want the Milwaukee using so much electricity that their other customers were left short.

Well, when Faudren asked me to make a fast run, I wanted to help him out so I agreed. Usually, engineers held their trains at either the bottom or the top of the hill, if they couldn't get over the hill before Peak Load time started. I was thinking I could make it over the hill before five and so I didn't stop.

Crew change at Cle Elum, Washington, 1961./Photographer unknown.

I was about halfway up the 2.2% grade out of Beverly when it started getting close to five. The substation operators knew I wasn't going to make it and they were getting worried so they started dumping the juice on me. I knew they were trying to make me stop, but I said to myself, "Why sit and wait for an hour when I can be over the summit by five minutes after five." When I didn't stop, they shut down Doris substation on me. I still kept going though. My motor was drawing enough current from Kittitas and Taunton to run even if Doris was shut down.

You were supposed to come down off the hill in regeneration, but I was a young guy and anxious to get in. When I reached the summit I decided to just let her roll. You could only go down the hill at 17 mph in regeneration but you could go 40 mph on air brakes. I got Faudren to Cle Elum on time but when I got there the dispatcher and all of the management were mad. They wanted to know what my excuse was. I told them I didn't have an excuse. I just wanted to get in.

Later I was called to a hearing about the matter in Othello. I figured they were going to rake me over the coals for not stopping and for not going into regeneration, but I thought they would have a hard time proving it. The first thing they did was bring out the substation ammeter charts. They pointed out on the graphs exactly what I was doing and when. Next they told me about all the problems I had caused for them. The Railroad's annual power rates with Washington Water Power were based on how much current the railroad used during Peak Load Time. As long as they didn't use any, they got the minimum rate. Because of my run they were in a high rate category for the entire year. Their first question for me was, "Were you the engineer of that train?" Second, they asked, "Is the engineer responsible for the safety and welfare of the train?" Third, "Were you conversant with the rules and did you obey the rules?"

They didn't take ignorance, insanity or loss of memory as excuses on the railroad. If you don't have a copy of the rules it is up to you to get them and learn them. I knew they had me so I admitted my guilt. They didn't lay me off, but I had to sign a form they gave me. I never would have run during peak load time if I had known I was going to cost them all that money for a year just because I wouldn't stop. They guys used to kid me about it and every once in awhile someone would come on the radio to ask if I was planning a little peak load running any time soon.

RIDING THE BEVERLY HELPER
By Don Dietrich

After moving to the West Coast, I wanted to take a ride on one of the Milwaukee Road's big electrics, and to ride a helper on one of the hills to see what goes on "mid-train." Both of these desires were filled quite by accident in October of 1966.

The weekend started as a trip to Yakima, Washington, to get some apples, with a railfanning extension to Pasco to photograph SP&S equipment. On the way back to Seattle, I decided to check in at Othello, the east end of the Milwaukee's Coast Division electrification.

When I arrived at the Othello yard, it appeared that westbound freight #263, powered by the E22ADB and two GP-9s, was ready to leave, The units used on the Coast Division were all boxcab units built by GE in 1915-16. They used 3000 volts DC and each unit weighed 144 tons. Most units were semi-permanently coupled in sets of four. The E22 was four units until the previous winter when the E22C was derailed and wrecked. It had not been replaced. The train started to move toward the depot where the train crew was waiting as I was looking the area over. I decided that I could not get very good photos here and thought that Beverly, a point 45 miles west would offer more attraction. So, I headed for Beverly.

The highway did not parallel the track from Othello to Beverly so I drove as fast as possible under the conditions. It was hunting season in Washington and to say the road was crowded is an understatement. The highway mileage is roughly 45 and the railroad made it in 37.8.

Beverly was a small community on the east bank of the Columbia River, some seven miles south of Interstate 90. It consisted of a cafe-store, service station and about 24 buildings in various stages of neglect. Indoor plumbing was not universally accepted and street paving and sidewalks were unheard of. Sagebrush and rocks outnumbered the residents 10,000 to 1. The reason Beverly existed at all was that at this point the Milwaukee Road crossed the Columbia River and began a 2.2% climb for 19 miles to cross the Saddle Mountains. A branch to the Atomic Energy Commission plant at Hanford swung off at the west end of the bridge and headed south. A small yard, wye and depot completed the rail facilities at Beverly.

What attracted the railfans to Beverly were the high bluffs on each side of the track which made for good photographs and the four unit helper stationed here for the westbound grade. E39ADCB was the helper that weekend. After taking photos of the helpers and a green Milwaukee Road reefer which was used as a company ice car, I waited for #263. Keeping an eye for the red block signal, I waited and waited. After about 90 minutes, I looked around and found the engineer for the helper washing the windows on his truck. He said that the running time from Othello was only 55 minutes, but that #263 was going to wait for a meet with #264 at Othello. That was holding him up.

The engineer continued that he was expecting #263 in about an hour. We got to talking railroading and he told me to go up in the cab and look around if I wanted. When he asked if I'd like to ride up the hill on the helper I snapped at the chance.

After another hour of waiting, the block turned red and a headlight came into view. #263 arrived in a cloud of brake shoe smoke, pulling to a halt short of the end of the bridge. The train consist included seven cars for Kittitas, which was the next station of any size, and the swing brakeman made a cut behind these cars. The head end pulled ahead and backed into the house track to pick up nine cars left by the Hanford patrol. These movements were all directed by the fireman on the helper who took the signals from the brakeman and relayed them to the train engineer by radio. The train was recoupled and #263 proceeded slowly across the bridge.

The train consist was 41 loads and 81 empties, 5247 tons. On the 2.2% grade west from

Beverly, boxcabs and GP9s were rated equally at 600 tons each. It was apparent that the 3 electrics and 2 diesels which were the road power could not handle the tonnage alone. As the train crept by slowly, the swing brakeman and helper conductor counted cars. The proper place for the helper was behind the 61st car. That turned out to be a flatcar load of machinery, so a cut was made between two box cars ahead of this point. The head end of the train proceeded slowly across the bridge, not coming to a halt until the engines were over halfway across. The helper now noiselessly entered the mainline from its storage track and backed onto the rear end of the train. Continuing almost noiselessly the helper pulled the rear end across the bridge until it coupled onto the standing head end. This placed the caboose right across from the Beverly depot and the conductor was able to alight, get his waybills for the pickup, and get back aboard. If there had been no pickup, the head end would only have pulled ahead far enough to let the helper in.

The swing brakeman made the airhose coupling in mid-bridge while standing on the catwalk, not an easy thing to do. He then gave the road engineer a radio highball, and set his air valve to give complete control to the road engineer. The fireman went back into the contactor compartment, threw a few switches and all hell seemed to break loose. The otherwise quiet cab was now filled with the noises of blowers, MG sets and other roaring equipment. After

E47 ACDB running as the Beverly Helper crosses the Columbia River Bridge at the foot of the grade./Photo by Robert Oestreich.

E33 ACDB sits beneath a boiling sun at Beverly. She was the helper of the day on September 2, 1962./Photo by Doug Cummings.

the fireman closed the contactor compartment door, things quieted down, but there was still a dull roar behind us.

The conductor in the caboose radioed a highball when the air pressure got up to par and the train engineer opened his throttle to get the train moving. Only after the train was rolling did the helper engineer open his throttle. He opened it gradually, watching the slack in the coupler on the car ahead to make sure he was pulling the cars behind, but not pushing the cars ahead. The throttle did not move easily and the engineer had to brace his foot against the front of the cab to move it. He opened it gradually, and after it was fully opened he did not touch it again until we reached Boylston at the top of the hill. Our speed was a constant 12 mph all the way. The engineer handled only the sanders, to prevent the wheels from slipping. He had to watch his field ammeter to check on his unit, but there were indicator lights for the other three units.

Sanding was a problem as the wind blows constantly in this area. Without the many cuts along the line where the sand stayed on the rails, the trains never would have made it up the hill.

Leaving the bridge, the Hanford branch swung to the south and the mainline swung north. As the train crawled up the hill there was a good view of the Columbia River and Wanapum Dam. On the left, old ties and some trolley still hanging were reminders of an old siding to a gravel pit. Five miles from Beverly, the line swung away from the river past the Doris substation. Here the voltmeter showed 3000 but at Beverly it was only 2800. It fell to 2200 by the time we reached Boylston. The overloaded substation equipment just could not keep the voltage up. In some cases this worked to an advantage, as the wheels slipped less on the reduced voltage.

The line continued in a series of S curves, making the head end of the train visible broadside except when it was hidden in a cut. The area is composed of rolling hills with lava outcroppings. Sagebrush and rattlesnakes were the only living population for most of the area. In one gully a spring must come near the surface as green grass and a few small trees broke the monotonous waste.

With the wind blowing through, the cabs got pretty cold in spite of foot warmers and a 3000 volt heater. Formerly there were small coal stoves as well in case of power failure, but they had been removed. The engineer talked of winter and the good snow photos that were available if you could get near the track. A 6 inch snowfall would produce 3 to 4 foot snowdrifts in the cuts because of the wind.

Ahead was Boylston tunnel, the top of the grade. The train passed through and the helper engineer radioed ahead when the helper was clear of the tunnel. The train stopped and the helper engineer attempted to bunch the slack ahead of him so that the swing brakeman could uncouple the helper. This was not as easy as it may sound since the head was on the downhill grade. There was a lot of free movement in the train even with the engine standing firm 60 cars ahead. At last he made the "pin" and the head end was pulled ahead to clear the crossover. The helper uncoupled from the rear end and backed through the crossover into the passing siding. The fireman then assisted the road engineer in putting his train back together, and #263 left for Tacoma.

As soon as the caboose cleared the east switch of the siding, the helper headed back to Beverly. As we entered the tunnel, the engineer put the units on regenerative braking.

The engine without train going through the tunnel sounded exactly like a set of subway cars going through the subway. Once clear of the tunnel, memories of riding the Illinois Terminal and the North Shore came to mind, the noises were so similar. It had grown dark outside and the headlight did its best to pick out the track ahead, but with one curve after another it was shining into space most of the time.

Our speed downhill was a constant 18 mph, with only two air applications to keep everything under control. The return trip was uneventful and we stopped at the east end of the bridge only long enough to open the switch into the storage track. The helper proceeded down toward the depot to register.

The crew had been on duty since 7 am, beginning its day by going over the hill light and helping #264 come back, staying in the train on the downhill side for braking power. The crew was on overtime when it helped #263 and was hoping that #261 would show up before its 16 hours were up so it could help them also. It was a typical day for the helper crew but it was a new experience for me.

When Don Dietrich rode the Beverly Helper, E22 ABD and two GP-9's were the head-end power. Here the set is shown westbound at Cle Elum, May 20, 1967./Photo by Doug Cummings.

A BLUE FLASH AT AUBURN

As told by Albert Farrow
Retired Northern Pacific Engineer

One night about 1950 or 1952 I was switching in the Northern Pacific yards at Auburn, Washington. It was around 3:00 a.m. and I had turned the operation of our diesel over to the fireman. I was just sitting on the left hand side of the cab looking out the window. The night was pretty clear with the moon out and all, and over on the Milwaukee line I could see one of their old boxcabs picking up some cars. They were about a half mile away and I could see them cut the cars in and do a brake test. They were getting ready to head out when all of the sudden there was a great big flash that lit up the sky. The boxcab just sat there looking like a fireworks display with cascading sparks pouring out of all the windows. It was kind of like Niagara Falls in sparks then suddenly everything went black. We figured anybody in that thing must be fried because nothing moved and their train just sat there blocking the highway. After about 20 or 30 minutes, our lunch break came and instead of going in for beans we jumped into the fireman's car and drove over to the Milwaukee line. The locomotive was silent and dark as we walked along the train, but when we reached the cab someone opened up a window and said hello. We were surprised those guys were alive so we asked them what had happened. They said that with 125 cars their locomotive was a bit overloaded. When they tried to start the train, they were drawing so much current that the trolley wire burned in half. One end of it dropped down on the roof of their boxcab and grounded out which made the big flash. They said they were waiting for a diesel to come out from Tacoma and haul them in. They said all of this kind of matter-of-factly like it was an everyday occurrence to them.

E50 AB with an eastbound freight passing Auburn depot, May 6, 1946./Photo by Al Farrow.

SHOEMAKER AT THE THROTTLE

As told by Ralph Edwards

For about a year during World War II I fired in passenger service for an engineer named George Shoemaker. Shoemaker was a guy who really knew a lot about Bipolars and he liked to run fast.

Our usual run was between Tacoma and Othello with a Bipolar and the second section of #18, the Columbian. The second section was day coaches and the first section was the sleeping cars.

A conservative engineer named Bill Jones was the engineer on the sleeper section. When he left Seattle, the only places where he had to pick up passengers were at Renton, Cle Elum, and Beverly. The stop at Beverly was for all of the people who were working at the Hanford Atomic Reservation.

On the second section we had a lot more stops. We picked up at Renton, Cedar Falls, Hyak, Cle Elum, Ellensburg, Beverly and any station in between if there was anything in the baggage car to be delivered.

On the way to Cedar Falls, me and Shoemaker would get the train heat boiler running real good so we wouldn't have to do anything with it for a while. When we left Cedar Falls, Shoemaker would tell me it was time for his lunch and a nap. Then he would let me run the Bipolar until we got to the top of the grade.

Coming out of Cedar Falls I had to lay back until the train ahead of us got to Garcia. By that time Bill Jones was running off the Hyak substation and it was all right to follow. Trains had to be spaced out on the hill to keep from pulling the line voltage down. If you pulled the voltage down too low, nobody could move.

After we got to Hyak it was downgrade and Shoemaker would take over again. Shoemaker would go like a striped ape. After each stop he would make up so much time that we were only one block behind the sleeper section. His favorite saying was, "That old sonovabitch can't leave me!" At Cle Elum we would catch him and at Beverly we would be right behind him.

As Bipolar Ten-Two-Fifty races south out of Seattle toward Black River, the fireman peers out a boiler room window. The NP-GN mainline is on the other side of the train. Date unknown./Photo by James A. Turner.

It was the same way on westbound trains between Seattle and Tacoma. The track was good for 70 mph and the Great Northern's passenger train to Portland left Seattle at the same time we left for Tacoma. After you passed Argo Tower, G.N. line ran parallel to the Milwaukee all the way to Black River Junction. The lines crossed at Black River and the last one there would get a red block from the interlocking tower.

Shoemaker would bring our train out and just kind of lay back waiting and tell me, "We aren't going to let that damn Great Northern beat us today." As soon as the G.N. came out it was a race. We would be side by side until we got close to Black River Junction and then the G.N. would pass us every time. The Renton substation operators had a hard time keeping the voltage up along there and when your voltage drops, your speed drops. Shoemaker would have a fit every time the voltage dropped down to 2400. When the G.N. went by, Shoemaker would cuss out the substation operators. Then he would tell me "If the Milwaukee just gave me a steam engine as big as the G.N. has, those guys would never catch me." Those big 2500's the G.N. ran were some powerful 4-8-2's and they could really move.

In Othello we had a roundhouse foreman named Schwanke. We called him, "Let her go Schwanke," because he always wanted to let the engines go out without his doing any work on them at the roundhouse.

One time Schwanke was in Seattle, and Shoemaker had been complaining to him about rough rides due to the lateral play in the driving boxes of the Ten-two-fifty-four (10254). On this particular night, Shoemaker invited Schwanke to ride in his cab with us over to Othello. He said to me, "I'll show that sonovabitch just how hard riding this thing is before we get to Othello." When we left Hyak, we were running a little late and he really wound her up. I can still see Schwanke sitting there trying to hang onto his seat with one hand and hold onto his straw hat with the other. When we stopped at Cle Elum, Schwanke told Shoemaker, "I'm going to get off this thing. I wouldn't ride it another inch, it'll beat you to death." He dismounted there and Shoemaker told him, "Now you know what we've been complaining about."

Schwanke didn't fix the driving boxes any better but I guess the shops were doing the best they could during the war. They just didn't have time. They had to keep everything moving.

G.N. #2500 leaving Seattle in 1938./ Photo from Warren Wing Collection.

THE FIRST TIME I FIRED ORPHAN ANNIE
By Bill Wilkerson

One night in 1946 I was called to fire No. 250, an oil burning steam engine. My run was from Miles City to Harlo on a coal burning district. Little did I know this was going to be a trip to remember nearly 40 years later.

The traveling engineer knew that I had fired oil on the Coast Division when I worked as a borrowed fireman the first three months of 1942. He called and told me that I probably knew more about firing oil than he did so he wouldn't go along that night. He would be needed more to instruct the fireman on mallet No. 57 when it was called. I assured him that while I had never fired Annie, all oil burners on the Milwaukee were the same and I was sure that I wouldn't have any trouble. I was never too fond of having an official along at any time, especially when he would be trying to instruct me on something he didn't know much about.

The enginemen and nearly everyone associated with the 250 called her Orphan Annie because for nearly eight years, she was the only 4-8-4 on the Milwaukee. She and mallet No. 57 had come to Miles City from the Idaho Division for Class IV repairs in the shop.

Orphan Annie would visit us every four years because all of her heavy repairs were done in the Miles City shop. Class IV meant a complete overhaul of the engine, including removal of fire tubes in the boiler. She would usually work in to Miles City on freight or run light. After she got out of the shop, she would usually run light back to the Idaho Division to break her in.

There was a coal miners' strike on in 1946, and the stock piles of coal were dwindling fast. The railroad decided to use Orphan Annie in passenger service between Miles City and Harlo until the strike was over.

I was firing the first section of the Olympian which was the sleeping cars. The second section was the day coaches. Because the sleepers were the heaviest, they usually had an S2 class 4-8-4. With Annie in passenger, there was one more S2 available for freight service. I was firing for Charles Nobel, who had lived in Harlo for years. He moved to Miles City when he could hold a passenger job.

Annie had sat around the roundhouse for about eight hours when we reported for duty about midnight. My oil was cold, so I turned on the tank and the line heaters. While they were heating up, I checked my oil and water supply in the tender. Unknown to me, it had been filled with highway road oil instead of locomotive fuel, but it all looked the same—black and gooey. I came back into the cab, lit some cotton waste and threw it into the firebox. Then I opened the atomizer a little and pulled the oil firing valve open until I heard the boom as the oil ignited. The steam pressure was down to about 150 lbs. and I allowed the fire brick to heat up slowly as we moved up to the depot to change engines.

By the time we were ready to leave town, I had the firebox hot and the full 230 lbs. of steam pressure. After we got out over Tongue River bridge (about a mile) and Charlie was picking up speed, I opened the peep hole in the firebox door and poured a couple of scoops of sand in slowly to cut out the soot. All our oil burners had a box for sand built onto the front of the oil tank. It held a couple of hundred pounds of sand. You had a small scoop that holds about a quart or so of sand. As you poured the sand slowly into the peep hole, the draft rushing in worked like a sand blaster to cut out the accumulated soot. You sanded leaving town while working a sharp exhaust until you got all the soot out and could run with a clear smoke stack. With a good hot firebox and a clean flame, soot would not build up very fast. Sitting around a terminal very long would build up a pretty heavy coating and this would cause the boiler not

Trans-Missouri Division Engineer Bill Wilkerson beside Harlowton Switcher E57B on 2-2-74. The Harlowton yard was part of the Trans-Missouri Division./ Photo from the Bill Wilkerson Collection.

*Orphan Annie, locomotive #250 outside of the Miles City shop at the completion of a major overhaul. Beside her are "all of the bosses" from the shop. Date probably 1942 or 1946./*Photo from the Bill Wilkerson Collection.

to steam freely. After I sanded her out and had a clean white fire, I sat down to enjoy the trip and watch the fire reflection on the ground. The front damper door was about four feet long, 12 inches high and hinged at the top, so it opened forward. This let the light from the fire reflect down on the ground ahead of your cab. At night, you could not see the smoke stack so you watched the color of this reflection to check on how clean your fire was burning. During daylight hours, you watched the amount of smoke coming out of the stack. Anything darker than a light brown was trouble and you had better start doing something.

Annie sounded beautiful at around 60 mph and rode like a coach, so I thought this was going to be a night to enjoy working. About 25 or 30 miles west of Miles City, I began having trouble keeping the steam pressure up to 230 lbs. and I noticed that my fire was flashing more yellow instead of white. I thought this was an indication of cold oil, although my temperature gauge indicated between 115 and 130 degrees, which was the proper temperature. I turned up the tank heater and turned the line heater on again. This didn't help, so I had Charlie drop the valve cut off lever down a couple of notches to sharpen the exhaust for more draft and I sanded it out again. Nothing made any difference—the black smoke continued to roll out the stack and the flame was a dirty yellow. Evidently there had been enough Bunker "C" oil in the bottom of the tank to get me 25 miles before the road oil started coming through.

I knew I was firing way too much oil, but any attempt to cut back caused me to lose steam pressure. Although Charlie was a 1910 engineer, he knew nothing about firing oil. In fact, he hadn't fired anything since 1910, and he was a hired engineer and didn't have a firing date on this division. He cooperated in running the engine any way I thought would help me, but nothing made any difference.

By now I had the hot oil so close to the flash point that it was exploding off the atomizer and it sounded like a machine gun in the firebox. I was able to keep from 200 to 220 lbs. of steam pressure so we could keep on time. That was all Charlie worried about, as the second section was only about ten minutes behind us and the line had no block signals. I had used all the sand by the time we were east of Ryegate, but we made Harlo on time.

When we cut off the train at Harlo, I looked into the firebox and it was all crusted up with a red glowing glass-like coating. It was built up so bad it was almost blocking the oil from the atomizer. The Harlo roundhouse spent all day chiseling it out of the firebox and boiler and I spent about half the day on the phone trying to explain to the master mechanic how I made such a mess. About that time, the crew of mallet No. 57 called in from Terry to say that they had died from low steam and their firebox was all crusted up with a red glass coating. After we checked what kind of oil we were burning, the Miles City store keeper confirmed that he had purchased road oil instead of Bunker "C." That was all he could get on short notice. Road oil isn't made to burn. It has additives in it to make it harden on the highway. I was finally off the hot seat, but for awhile, the officials sure wanted my scalp or any other part of me they could get. Actually, I should have been complimented. My unorthodox method had gotten me 217 miles without any delay. With the official instructing the fireman on the 57 according to the book, they couldn't make it 40 miles.

Harlo had Bunker "C" oil, so we had a very enjoyable trip going home. I fired Annie several trips before I got bumped off the job, and they were all very enjoyable trips. She performed perfectly with the right oil and I hated to lose the job. Firing oil was always fun for me.

About four years later, I caught her as an extra engineer when she was returning to the Idaho Division after going through the Miles City shop. It was only a break-in run to Melstone. We had to stop and check everything every few miles to make sure nothing was running hot or cutting, but I was the engineer on Annie for 112 miles. It doesn't qualify me as an expert engineer or authority on Annie, but it at least got me into the club of legions of men who ran Annie and I'm happy about that.

E45B
E45B

E41 ACB in original black paint scheme. E41C has an oxide red roof. Some, but not all, Milwaukee electrics and steam locomotives had red roofs. Date and location unknown./Photo by Harry Morgan.

E81 switching at Deer Lodge in the Summer of 1972./Photo by Noel T. Holley.

Opposite Page: Bitterroot Mountain helper E45 awaits a train at Haugan, Montana, September 1972./Photo by Noel T. Holley.

E71 and E79 at Harlowton. Date unknown./Photo by Harry Morgan.

A brakeman rides the front platform of E79 as it passes the Continental Divide marker on Pipestone Pass. A 68 car train pulled by E21 and E73 hold the main and wait to enter Pipestone Tunnel. March 11, 1973./Photo by Rick A. Zeutschel.

Westbound E75 as seen from the cab of another Little Joe. Three Forks, Montana, March 1973./Photo by Noel T. Holley.

Milwaukee Road Little Joes E72 and E79 eastbound out of Anaconda, August 10, 1973./Photo by Casey Adams.

E78 on the Avery ready track, the morning after my electric journey. March 1973./Photo by Noel T. Holley.

E70, E79 and E73 wait on the Deer Lodge ready tracks on March 17, 1973. E47 has been partially dismantled and will never run again./Photo by D. Larry Zeutschel.

Westinghouse Motor E11 at Avery, Idaho. Date unknown./Photo by Harry Morgan.

E21 with the Olympian Hiawatha near Eagle's Nest Tunnel. Date unknown./Photo by Barry Kirk.

Passenger Motor E23 with a train at the Ellensburg, Washington depot./Photo from Bruce Butler Collection.

Freight Motor E25 eastbound over Interstate 90 on the Rye Creek trestle in the Saddle Mountains. December 1970./Photo by Gil Hulin.

Freight Motor E25 eastbound near Kittitas, Washington. December 1970./Photo by Gil Hulin.

Bipolar E4 at Black River Junction, Washington, May 1955./Photo by Warren Wing.

Freight Motor E50 nears Othello, Washington, April 1971./Photo by Gil Hulin.

Two freight motors E33 and E39 sit beside the Othello sand tower./Photo by Bruce Butler.

Telephoto shot of Three Forks Yard shortly after a snow storm. March 1973./Photo by Noel T. Holley.

#1201 an Idaho Division "C Engine" 2-8-0. Date and location unknown./Photo by Harry Morgan.

Bipolar E5 at Othello./Photo by Harry Morgan.

Yard Goat E82 at Deer Lodge./Photo by Harry Morgan.

Little Joe E21 and Passenger motor E22 in Deer Lodge, March 3, 1956./Photo by Bruce Butler.

Little Joe E21 is seen from the rear as she leaves her train in Deer Lodge, March 3, 1956./Photo by Bruce Butler.

Bipolar E5 at Othello./Photo by Harry Morgan.

Bipolar E3 with a troublesome train heat boiler, probably passing Taunton, Washington./Photo by Harry Morgan.

Bipolar E1 prepares to leave Seattle Union Station for Tacoma./Photo by Harry Morgan.

Bipolar E1 probably at Othello./Photo by Harry Morgan.

Westinghouse Motor E10 at Deer Park, Montana./Photo by Barry Kirk.

One of the first group of road/rail trolley repair trucks./Photo by Harry Morgan.

Bipolar E4 at Deer Lodge./Photo by Barry Kirk.

Little Joe E20 with the Olympian Hiawatha in Montana./Photo by Laurence Wylie.

E45 in helper service at Penfield, Montana July 28, 1973./Photo by Rick Zeutschel.

E45 in head-end helper service at Janney, Montana, May 12, 1973./Photo by Rick Zeutschel.

E45 couples on to the point of a diesel powered freight at Butte yard./Photo by Rick Zeutschel.

Train #264 at Kohrs, Montana May 5, 1974./Photo by Larry Zeutschel.

Train #263 on Donald Trestle, Donald Montana May 18, 1974./Photo by Rick Zeutschel.

Train #263 at Donald, Montana May 18, 1974./Photo by Larry Zeutschel.

Train #263 at Newcomb, Montana. E21, E78, 4005 and 19 are the motive power. E73 is in tow. June 1, 1974./Photo by Rick Zeutschel.

Little Joe E70 on display at Deer Lodge, Montana March 15, 1975./Photo by Larry Zeutschel.

Crew change at Alberton, Montana, June 2, 1974./Photo by Keith Newsom.

Milwaukee right-of-way in the Cascade Mountains near Garcia, Washington. The heavy rains and snows of western Washington kept everything here green. March 1961./Photo by Warren Wing.

TECHNICAL APPENDIX

TYPES OF LOCOMOTIVES

On the Milwaukee, the electric locomotives were generally known as "motors" rather than engines, the reason being that engines generate their own energy and motors do not. A steam engine is a rolling steam generator mounted on top of drive gear which converts the steam into tractive power for pulling trains. A diesel-electric engine is a rolling diesel driven electric generator, mounted on top of drive gear which converts electricity into tractive power. By contrast, an electric motor only converts electricity into tractive power. The generator is located elsewhere in a dam or a steam plant. The equipment mounted above the running gear of an electric motor is simply the equipment for controlling the traction motors in the drive gear.

CLASSIFICATION OF WHEEL ARRANGEMENTS

The wheel arrangement of electric and diesel locomotives is described using a series of letters and numbers. The letters describe power axles and the numbers describe unpowered axles. If there are no dashes or plus signs between the letters and numbers, then all axles are in one rigid frame or locomotive truck. Dashes between letters or numbers describe the use of separate trucks which pivot on a shared frame. Plus signs describe frames or truck which are connected through articulated joints, drawbars or couplers.

Using this international method of description, a 4-8-4, Northern type locomotive would be described as a 2-D-2. "D" is the fourth letter in the alphabet and describes four powered axles which have no unpowered axles between them.

The reason for classifying electric and diesel locomotives by this system is to avoid confusion about what the function of their wheels or axles is. On a steam engine it is reasonable to assume that the first and last sets of wheel are idlers, used for guiding the locomotive. The middle numbers are, of course, drivers. Under the Whyte classification system, which describes wheels, 2-6-6-2 is thus quite clear in describing the articulated steam engines the Milwaukee used on its Idaho Division. It is less clear in describing the Bipolars, however. Bipolars had a 6-8-8-6 wheel arrangement, but the leading and trailing wheels were the only idlers. Thus these passenger motors are classified as 1B + D + D + B1. The Fairbanks-Morse Erie-built diesels which once hauled the Olympian Hiawatha were 0-6-6-0's, as were the SD7's. However, a difference between the two is apparent under the system of letters and numbers. The Erie-Builts were an A1A-A1A type while the SD7's were a C-C. The Erie-Builts had an idler axle in the center of each truck. On the SD7's all axles were powered.

The five types of electric locomotives on the Milwaukee had wheel arrangements as follows:

GE Switcher, B-B

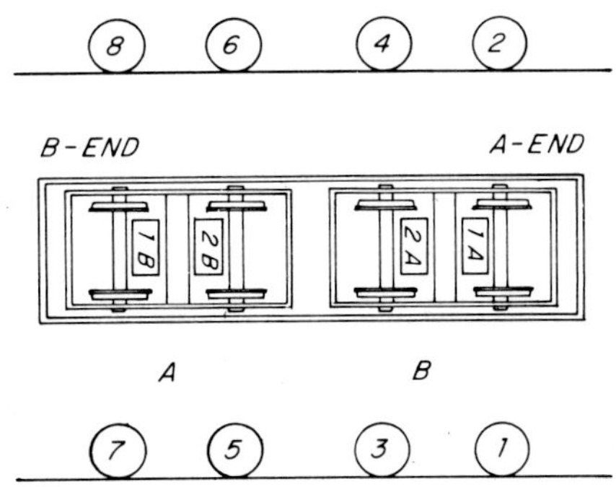

GE Bipolar, 1B + D + D + B1

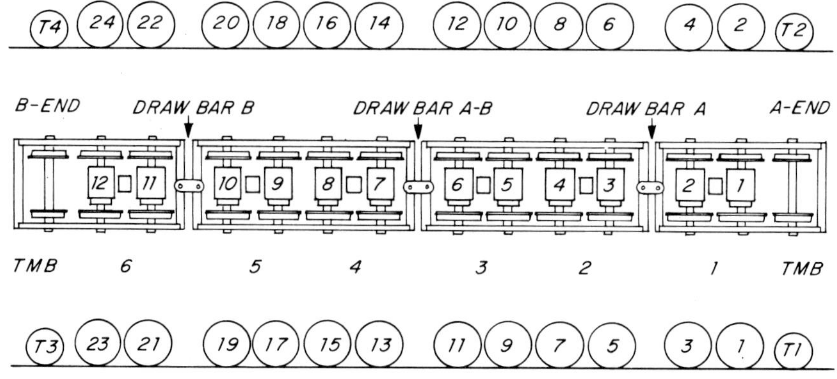

Westinghouse Motor, 2-C-1 + 1-C-2

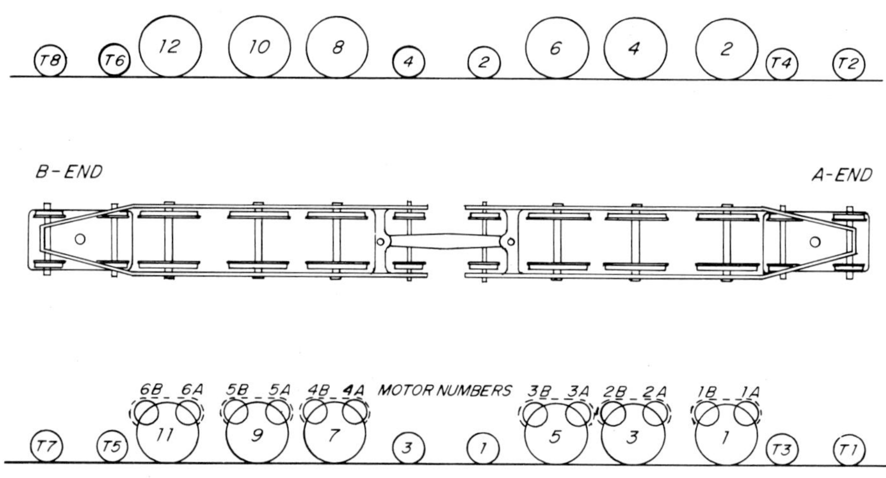

GE Motor, 2-B + B + B + B-2

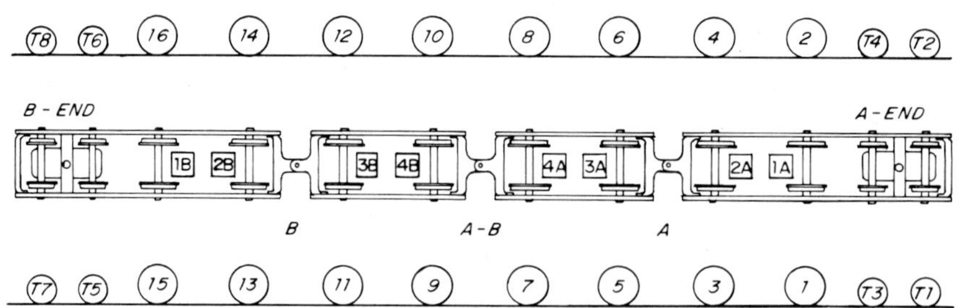

GE Little Joe, 2-D + D-2

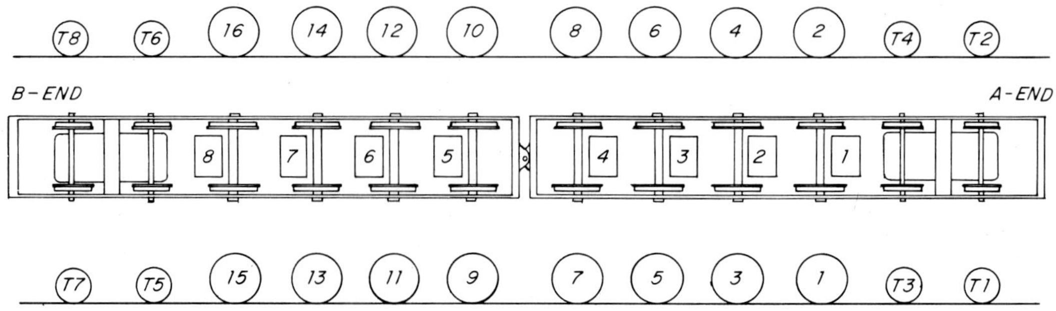

LOCOMOTIVE FRAMES

Mainframes of the Milwaukee's electric road locomotives were the driving wheel truck frames. Beneath each locomotive unit these truck frames were connected together by drawbars. Frames at each end of a locomotive unit were connected together by couplers or drawbars. All pulling and pushing forces were carried by the truck frames while the locomotive body simply rode above the trucks.

The Milwaukee's steeplecab switchers as well as most diesels and electrics in use today utilize a different design. The locomotive body contains the mainframe. Couplers are attached to the body and the trucks pivot independently beneath the body.

The two primary reasons for articulating the trucks and using them as mainframes are to obtain ruggedness and good tracking in a locomotive. Truck frames must be built ruggedly regardless of how they are attached to the rest of the locomotive. If they are linked by drawbars mounted at coupler level they form a very rugged chain. Pulling and pushing forces pass through the locomotive in nearly a straight line. Designing the trucks to link as a chain also makes them track better because they guide each other.

A rigid frame locomotive such as a steam engine tends to nose from side to side as it goes down the track. This tendency is reduced and sometimes eliminated by using guiding trucks to help keep the locomotive centered. Even with the use of guiding trucks, a large locomotive can exert severe impacts on the rails if operated at high speeds. During the development of its famed GG1 electric, the Pennsylvania Railroad found a 2-C + C-2 wheel arrangement to be much easier on the track that a 2-D-2 of equal size and power.

Independently pivoting trucks also tend to nose or oscillate. This too can cause problems. When Amtrak's E60C's entered high speed service in the Northeast Corridor it was found that powerful truck oscillations were knocking over rails and causing derailments. Locomotive mounted shock absorbers were used to prevent the oscillations.

Like the GG1, the Milwaukee's electric road locomotives never experienced serious problems with nosing or truck oscillations. Since the trucks were linked together they worked against each other to prevent such problems. For instance, in order for the front truck to nose to the right, it must force the truck behind it to simultaneously nose to the left. The Bipolars with their 1B + D + D + B1 wheel arrangement carried the philosophy of anti-nosing, anti-oscillation design further than any other Milwaukee locomotive.

DIRECT CURRENT DRIVE

Electrically, the locomotives owned by the Milwaukee were similar to the streetcars which once dotted the United States. They drew current from the overhead wires and used it in their motors without requiring the use of on-board transformers, motor-generators, or rectifiers.

A direct current electrical system was chosen because high-horsepower variable speed DC traction motors are easy to build and maintain. Alternating current presents a number of problems for use in locomotives. Commercial frequency 60hz AC does not work well in variable speed motors. In order to utilize AC, the railroads have been forced to either convert its frequency to 25hz in substations, or convert the AC to DC on board the locomotives.

Simplicity in electrical design was an important consideration when the Milwaukee chose to electrify as a DC system. The current in the trolley wires was current that the traction motors could run on. 3,000 volts flowed to the locomotives through the overhead wires. It was fed through groups of traction motors, and then returned to its source through the rails.

TYPES OF TRACTION MOTORS AND DRIVE SYSTEMS USED ON THE MILWAUKEE ROAD

Axle Hung Traction Motors

Axle hung, nose suspended traction motors were used on the GE Motors, Steeplecabs, and the Little Joes. This is also the type of traction motor used on diesel locomotives. A high speed electric motor with a small pinion gear on its drive shaft turned a large reduction gear on a driving axle. The motor housing was clamped to the driving axle. This kept the gears in proper alignment. The majority of the motor's weight was, however, suspended on the truck frame. A major advantage of this system was that traction motor weight and center of gravity were kept low to the rails. Additionally, since most of the motor weight was supported by the truck frames, the motor and track were cushioned by the springs in the truck. A disadvantage seen on the GE Motors, Steeplecabs, and diesels was mechanical weight transfer during acceleration. The weight transfer tended to cause wheel slip and could delay or stall the locomotive. The transfer occurred because the motors within a truck were hung in opposite directions. While making the locomotive move forward they turned in opposite directions. As they interacted with the gear train, one motor tended to push its axle down onto the rail while the motor opposite from it tended to lift its axle up away from the rail. With a lack of weight or pressure on it, one wheel set in the truck tended to slip and spin.

When motors are connected in series, the motor that turns fastest will use the most current. The Milwaukee electric locomotives started trains with all motors in series. This meant that if wheel slip occurred, the locomotive could stall. The free spinning motor could draw enough current away from the others to prevent them from turning.

The Little Joes did not have this mechanical weight shift problem because all motors within each truck were hung in the same direction. It causes no significant weight transfer if all motors in a truck try to lift up their axles at once.

Bipolar Traction Motors

In a bipolar drive system, large slow speed motors are used to power the locomotive. The traction motors are in three separate parts. The armature is carried on the axle and the axle itself is the motor drive shaft; thus no gearing is involved in the system. The field poles are attached to the locomotive frame. Only two poles are used in this system and they are mounted across from each other. The armature lies between them. A major advantage which this drive system had over that used in the GE Motors was that it avoided the mechanical weight transfer which could cause wheel slip. The weight and power of the traction motors was concentrated directly over the driving wheels. There was no way for motor torque to lift wheels off the rails. Additionally, this drive system is slightly more efficient than one which uses gears because there is no horsepower lost in turning gears. It is also much quieter than geared systems. Although the Milwaukee's other electric locomotives did not make much noise, much of the noise they made was gear tooth growl.

Drawbacks involved with the bipolar drive system are: 1) an axle mounted armature is not cushioned from jolts on rough track and the rails can be hammered by this extra heavy axle; 2) the locomotive cannot be regeared for different types of service; 3) since the axle mounted armature can move up and down in the locomotive chassis while the locomotive travels down the track, it does not always have the best alignment with the field poles.

The bipolar drive system has seldom been used on new locomotives since the time the Milwaukee Road purchased its EP-2's. This is because the design was made technologically obsolete by the development of improved, small, high speed, high horsepower traction motors that are of the axle hung, nose suspended type.

Twin Armature and Quill

The traction motors used on the EP-3's which the Milwaukee purchased from Westinghouse, were a twin armature with each armature geared to a quill. Each of the 1,500 volt traction motors was mounted directly above an axle and contained two 750 volt armatures within its housing. These armatures were wired in series so that the operating voltage of the motor was 1,500. Each traction motor had four field poles, also contained within the housing. The reason which Westinghouse gave for designing twin armature motors was improvement of commutator operation. Low voltage commutators generally cause fewer electrical problems than high voltage commutators.

The quill drive was a spring cushioning system designed to prevent the traction motors from suddenly applying full power and torque to the wheels. The quill was a hollow steel tube around the axle. The large reduction gear was mounted on the quill rather than the axle. The quill had a "spider" with seven spokes at each end. These were connected to the seven spokes of the driving wheels through coil springs. Sudden starts would compress the coil springs which would then cushion the jolt of sudden starts.

A major advantage of this system used by Westinghouse was that it avoided wheel slip due to mechanical weight transfer. Since the traction motors were above the wheels instead of beside them, their torque did not tend to lift the wheels or press them down. The torque was exerted laterally instead of vertically. The driving axles could not shift laterally within the frame, thus all torque from the traction motors went into turning the wheels.

In service, the Milwaukee found the spring cushioned part of the drive systems to be unnecessary. What is more, the difficulty and expense of maintaining such drive systems made the quill drive a disadvantage in a locomotive.

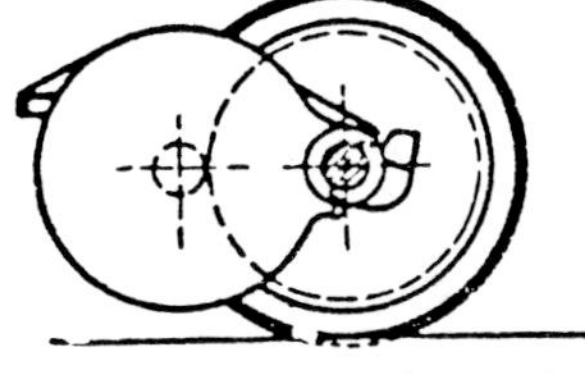

Axle Hung, Nose Suspended Motors, Single Reduction Gear.

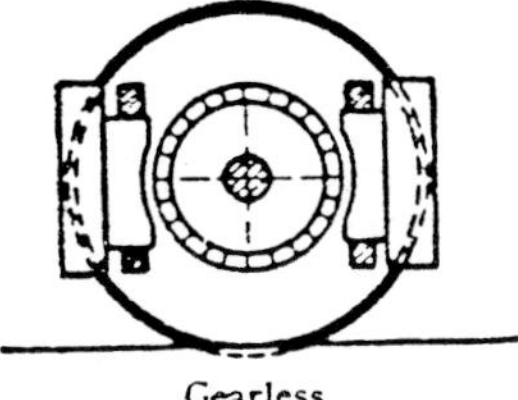

Gearless. Armature on Axle.

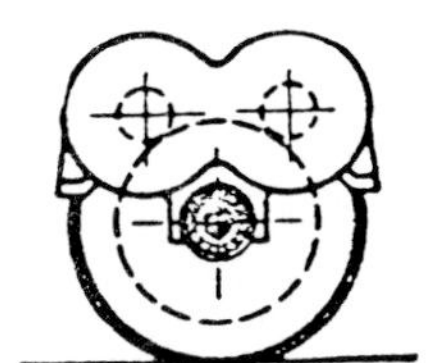

Motors and Quill on Frame.

TRACTION MOTOR COMBINATIONS

The traction motor combinations on each type of locomotive are as follows. In equating them with speed consider that a 1,500 volt motor receiving 1,500 volts will operate at full speed. If it receives half of that voltage (750 volts) it will run at approximately half speed, etc.

GE Motors and Steeplecab Switchers

Four 1,500 volt traction motors on each unit.

SERIES COMBINATION:	4 motors in series, 750 volts to each motor
PARALLEL COMBINATION:	2 groups of 2 motors in series, 1,500 volts to each motor

GE Bipolars

Twelve 1,000 volt traction motors in each locomotive

SERIES COMBINATION:	12 motors in series, 250 volts to each motor
SERIES-PARALLEL COMBINATION:	2 groups of 6 motors in series, 500 volts to each motor
LOW SPEED PARALLEL COMBINATION:	3 groups of 4 motors in series, 750 volts to each motor
HIGH SPEED PARALLEL COMBINATION:	4 groups of 3 motors in series, 1,000 volts to each motor

Westinghouse Motors

Six 1,500 volt traction motors in each locomotive

SERIES COMBINATION:	6 motors in series, 500 volts to each motor
SERIES-PARALLEL COMBINATION:	2 groups of 3 motors in series, 1,000 volts to each motor
PARALLEL COMBINATION:	3 groups of 2 motors in series, 1,500 volts to each motor

GE Little Joes

Eight 1,500 volt traction motors in each locomotive

SERIES COMBINATION:	8 motors in series, 375 volts to each motor
SERIES-PARALLEL COMBINATION:	2 groups of 4 motors in series, 750 volts to each motor
PARALLEL COMBINATION:	4 groups of 2 motors in series, 1,500 volts to each motor

TRACTION MOTOR SHUNTING

Full tractive effort was produced by the traction motors when all of the current fed to the traction motors flowed through both the armature and the field coils before going to ground. If, however, full tractive effort was not needed to pull the train, it was possible to shunt the tractive motors and produce more speed. A shunt weakened the field strength of the traction motors allowing a portion of the current to bypass the field coils. A strong magnetic field in the field coils gave strength to the rotation of the armature but also tended to hold back the speed of rotation. A weak field weakened the strength of rotation but allowed faster rotation. Thus while operating in a shunt, the locomotives produced less tractive effort than they otherwise would have but they achieved their highest speeds. Shunts were generally used on level track or with light train loads. The Bipolars used a field tap rather than a shunt. The field tap weakened the field strength by reducing the number of field coils in use. If 50% of the field coils were used, the result was 50% field strength.

THROTTLE OPERATION

Certain notches on the throttle quadrant were running positions. These determined speed and power by setting up different traction motor combinations. The Milwaukee's electrics produced their greatest pulling power with their traction motors connected in series. The greatest speed was possible while running in parallel.

Throttle settings did not equate directly with locomotive speed. The speed achieved at particular throttle settings varied. It was affected by train weight, track gradient and line voltage in the trolley.

When controlling a train, the engineer was limited to only a few choices for train speed. For instance, a GE Freight Motor with an average train on relatively level track might run at 15 mph in a series, 30 mph in parallel and 35 mph in the parallel shunt. The engineer did not have the option of running continuously at 20 mph. That did not match up with any traction motor combination and thus there was no running position on the throttle for that speed. All of the throttle notches between running positions were resistance notches. These connected various numbers of resistor grids in series with the motors, and provided for a smooth transition between motor combinations. Each resistor group could only handle the high amperage loads of train operation for two or three minutes. If the engineer were to use a resistance notch for continuous operation, he could cause an electrical fire as the iron resistor grids overheated, melted and burned.

THROTTLE POSITIONS ARE AS FOLLOWS:

GE Motors:	1st - 16th notches	series resistance
	17th notch	full series, running position
	18th - 19th notches	blank
	20th - 30th notches	parallel resistance
	31st notch	full parallel, running position
	32nd notch	shunt, 47% field strength
Bipolar:	1st - 9th notches	series resistance
	10th notch	full series, running position
	11th - 16th notches	low speed parallel, resistance
	17th notch	low speed parallel, running position
	18th - 25th notches	high speed parallel resistance
	26th notch	high speed parallel, running position
	Shunt separately controlled, 50% field strength	
Westinghouse Motor:	1st - 13th notches	resistance
	14th notch	running position, full field
	15th - 16th notches	shunt, 86% field strength
	17th notch	shunt, 72% field strength
	Motor combinations separately controlled	
Little Joe:	1st - 15th notches	series resistance
	16th notch	full series, running position
	17th - 25th notches	series parallel resistance
	26th notch	series parallel running position
	27th - 36th notches	parallel resistance
	37th notch	parallel, running position
	Shunts separately controlled, 74% and 54% field strength	

REGENERATIVE BRAKING

When an electric locomotive is "motoring" it consumes current from the railroad's electrical distribution system. The current turns the traction motors and the locomotive pulls the train.

If a train is slowing down or descending a mountain grade the train will push the locomotive. With gravity and train weight causing the traction motors to turn, the motors can be operated as generators. The value in using them as generators is that it requires large amounts of energy to turn generators. When used as generators, the traction motors will hold back or reduce the speed of the train. This makes it possible to reduce the use of train air brakes. Heavy use of air brakes while a train is descending a long grade produces hot wheels, brake smoke and sometimes cracked wheels. Car wheels can crack when the friction of iron brakeshoes on steel wheels causes the wheels to overheat.

In regenerative braking the locomotive feeds current back into the electrical distribution system. Braking power begins when the voltage produced by the locomotive exceeds the trolley wire voltage. For instance, Little Joes often fed direct current at 3,400 volts into a trolley wire carrying 3,200 volts.

On the Milwaukee Road, regenerated current could be used to power other trains if any were nearby. If none were nearby, it would flow back through the substation to the commercial power grid and deduct itself off the railroad's electrical bill in the process.

The directly measurable economic value of regeneration was at times very significant. The eastbound grade across the Rockies was 1.66% for 10 miles from Butte to the summit of Pipestone Pass. Descending, it was 2% for 27 miles to Piedmont. Because the descending grade was longer and steeper than the ascending grade, an eastbound train could generate more electricity than it consumed while crossing the Rockies. The power bill would thus be less than zero, with Montana Power owing the railroad money for the trip. In terms of systemwide costs, regeneration reduced the Milwaukee's electrical bill by 12% annually.

Dynamic braking is similar to regenerative braking. It is used on diesels and also on some electric locomotives. An example is the E-44 electric which the Pennsylvania Railroad bought from General Electric. Dynamic braking differs from regenerative braking primarily in that the regenerated current is fed into resistor grids onboard the locomotive and dissipated as heat. Some electric railroads operate under contracts where the power company will not pay for or accept electricity generated by customers. Some direct current lines have substations which cannot convert DC back into AC. If the current goes nowhere, trolley system voltage will rise to dangerous levels.

Some electric railroads have thus been faced with the choice of putting resistor grids on the locomotives, putting them in the substations, or not using the traction motors as brakes. Diesels of course have no trolley wire to feed the current back into. Thus they use onboard resistor grids.

PANTOGRAPHS

Milwaukee Road pantographs were designed to handle high levels of current and a wide range of trolley wire heights.

The Milwaukee's standard trolley wire height was 24 feet 2 inches above the rails. It could, however, go as low as 18 feet 9 inches when passing through tunnels and bridges. In order to accomodate any reasonable variation from the standards, the pantographs could function properly as high as 26 feet and as low as 17 feet above the rails.

Large DC locomotives can draw tremendous amounts of current and each Milwaukee Road pantograph could carry as much as 1,200 amps. In order to do this without an excessive amount of arcing, each pantograph had two shoes. The shoes were made of thin pressed steel and each had two copper wearing strips on it. The copper was lubricated with a mixture of axle grease and graphite. This lubrication kept trolley wire wear extremely low, reduced the amount of arcing and prevented friction from causing the wire and pantograph to sing. The pantograph frame was made of lightweight tubing and had joints made of brittle aluminum. Cross bracing prevented side sway at high speeds. The pantographs were held in contact with the wire by a set of springs. In order to raise the pans, air cylinders drew the pan springs tight. To lower the pans the air was released and the slack springs allowed the pan to drop.

Pantographs on the Milwaukee were narrower at the top than they were at their middle or bottom. They were designed this way to provide for better clearance through tunnels and bridges. The frame tapered toward the top, and the flat section of the pantograph shoes was only 2 feet 10 inches across. The downward sloping part of the pan shoes, called "horns", were unusually long and stretched out 6 feet 8 inches from tip to tip. Pantographs on other railroads tended to be straight up and down, they had broad flat pan shoes 4 to 6 feet across and the pan shoes had short horns.

Each road locomotive had at least two pantographs. Under most conditions only one would be used at a time so that the other could act as a spare in case of damage. If the locomotive was drawing more than 1,000 amps or if frost, sleet or dirt on the trolley wire caused excessive arcing, both were raised. When only one was in use, the rule books suggested using the front pan on all but the Bipolars. On a Bipolar, the pans were mounted so close together that if the front one were wrecked and torn from the locomotive, it could take the rear pan with it.

Many things could cause a wrecked pantograph. Damaged trolley wires could snag it, or a badly worn pan shoe could tear loose and take parts of the pantograph with it. One common problem involved the pan horns rising above the trolley wire and hooking the wire hangers. This generally occurred when a locomotive was leaving a siding which had slack trolley. The trolley above the siding might rise as high as 25 or 26 feet due to pantograph pressure. However, the mainline would remain at the standard 24 feet. As the pantograph approached a convergence of the wires above a switch the pantograph horns might be higher than the mainline trolley wire. The pan horns would then just hook the mainline wire rather than riding under it. In order to prevent these types of problems the Milwaukee installed "bows" on most of its pantographs. With these devices on the pan, the horns were prevented from hooking the wires and the pan was more likely to move safely from slack wires to taut ones. The Great Northern and the Virginian also used bows to prevent these problems. The Pennsylvania, the New Haven and most of the European railroads maintained their trolley wires in such precise alignment that pantograph bows were unnecessary. Most railroads construct their pantographs of weak and brittle materials. This is to make sure that whenever pantographs get tangled in the trolley wires, the pantograph will tear loose from the locomotive rather than ripping down sections of the trolley.

Losing all pantographs meant losing all power and was a situation train crews dreaded. One dismal winter day in the 1970's when Alan Burns worked as a Milwaukee Road brakeman, he witnessed such an event. A train pulled by two Little Joes was dropping down the Bitterroot grade toward Avery with all pans raised. Frost and heavy regeneration meant that all four were needed. Arc flashes marked the train's descent into the valley when suddenly, like dominoes, all four pantographs tore loose one after another. Crumpled and broken, they dropped into the snow as a 10,000 horsepower locomotive team ground to a halt. The entire train just sat until diesels arrived from Avery to provide braking power and safely bring it into the terminal.

The Milwaukee manufactured pantographs in its shops at Tacoma and Deer Lodge. They were stockpiled in all engine terminals to allow for quick repairs. In 1972, the Milwaukee estimated the value of each pantograph at $1,500.

The single arm Faiveley and Fairley pantographs common in other countries were never used on the Milwaukee. They were never needed. Those lightweight pans were originally developed to provide better pantograph tracking on lines where average train speeds were above 60 mph and the trolley wire height was fairly constant. The Milwaukee did not operate at speeds above 60 mph for long distances. Additionally, the Milwaukee's trolley wire height variations were extreme by European standards. A single arm pantograph which is large enough to accomodate those variations requires a number of modifications in order to provide improved high speed tracking.

The Pennsylvania Railroad initially adopted Faiveleys because of their compactness rather than their high speed tracking abilities. The Pennsy's E-44 rectifiers purchased in the 1960's only had room for one standard pantograph on each locomotive. For safety reasons, the Pennsy wanted two pans on each locomotive. It found that this would be possible if two Faiveleys were mounted back-to-back above the cab.

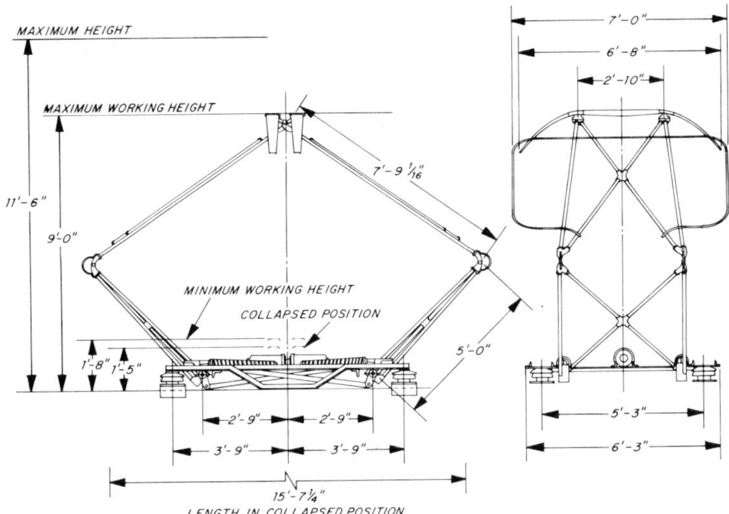

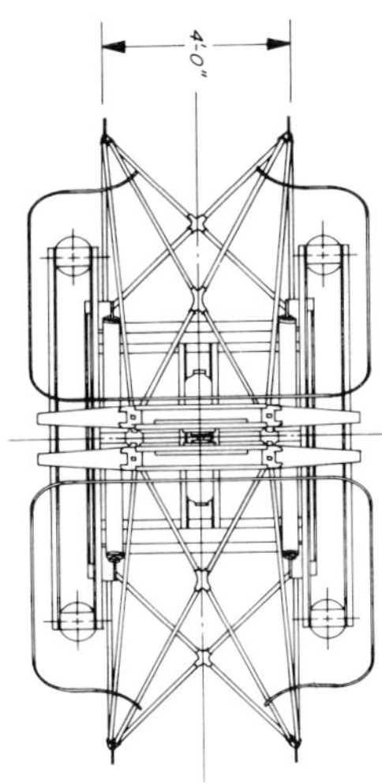

ELECTRIC LOCOMOTIVE SPECIFICATIONS

Many different publications have listed differing statistics on the weight and performance of Milwaukee Road electrics. This book quotes statistics which the author believes to be correct, however no claim is made that the data is indisputable. Indisputable statistics may not even exist, since the Milwaukee made no great effort to produce them.

An example of the difficulty in verifying data is the case of official weight for a Westinghouse Motor. The 1926 employees timetable for the Rocky Mountain Division lists them at 307 tons. The 1926 Coast Division timetable lists them at 276 tons. The operator's manual dated 1926 lists them at 310 tons and implies that motors with Delta trailing trucks may weigh more than that. The manual does, however, point out that all weights are approximate since no locomotive scales existed in electrified territory.

LOCOMOTIVE CLASS DESIGNATIONS

Locomotive Class	Locomotive Type
• ELECTRIC FREIGHT MOTORS	
EF-1	Two unit GE Motor set.
EF-2	Three unit Motor set. Full-size middle unit.
EF-3	Three unit Motor set. Short middle unit.
EF-4	Little Joe.
EF-5	Four unit Motor set. Middle units may be long or short.
• ELECTRIC PASSENGER MOTORS	
EP-1	Two unit GE Motor sets with high speed gearing and train heat boilers. Operated 1915 to 1920.
EP-1A	GE Motors rebuilt for passenger service in the 1950's. One unit in each set had a rounded front.
EP-2	Bipolar.
EP-3	Westinghouse Motor.
EP-4	Little Joe equipped with roller bearings and train heat boiler.
• ELECTRIC SWITCHING MOTORS	
ESO-1	Company built single truck shop switchers. Used only at Deer Lodge roundhouse and operated on extension cords.
ES-1	GE Steeplecab used in Great Falls (1500 volts).
ES-2	GE Steeplecab (3000 volts).
ES-3	GE Motors. Same as freight units but equipped with footboard pilots and headlights front and rear. May be designated EF-1 "half unit". ES-3 may be single cab unit or cab plus short unit.

Various figures have been published as the maximum speeds of Milwaukee Road electrics. Part of the reason for this is the existence of several types of maximum speeds, and also the impact of design changes.

"Maximum Safe Speeds" are those beyond which the manufacturer or the locomotive shop cannot guarantee the proper functioning of the locomotive. The mechanical limit is the speed which the locomotive does not have the power to exceed. Another type of maximum is the speed limit which may be applied to a class of locomotives if mechanical problems tend to be related to higher speed.

GE Freight Motor

Original maximum safe speed was 35 mph. At higher speeds centrifugal force could tear the armatures apart. Maximum speed was raised to 45 mph in the 1950's. Rebuilding included high strength steel banding to protect the armatures from coming apart, and additional traction motor shunting.

GE Passenger Motor

Original maximum safe speed was 70 mph. When rebuilt for passsenger service in the 1950's, E22 and E23 were rated at 75 mph. This resulted from additional traction motor shunting, and steel wire banding on the armatures.

GE Switcher

Maximum safe speed was 35 mph. Speed was not necessary for switching.

GE Bipolar

According to General Electric, the maximum safe speed was 90 mph. Maximum speed allowed by the Milwaukee's rule book was 60 mph. As a result of rebuilding in 1953, the maximum was raised to 74 mph. After assignment to the Rocky Mountain Division they were operated up to 85 mph, but mechanical and electrical problems forced the imposition of a 60 mph limit.

Westinghouse Motor

Maximum safe speed according to the rule book was 60 mph. According to train crews and retired officials, these passenger motors were commonly operated at speeds above 80 mph.

GE Little Joe

Maximum safe speed with plain bearings 65 mph. Maximum safe speed with roller bearings 70 mph. Maximum speedometer reading 84 mph.

LOCOMOTIVE PERFORMANCE STATISTICS

CLASS	TOTAL WEIGHT (POUNDS)	WEIGHT ON DRIVERS (POUNDS)	TRACTIVE EFFORT AT 18% ADHESION	TRACTION AMPS CONT.	MOTOR HOURLY	HORSEPOWER CONT.	HORSEPOWER HOURLY	TRACTIVE-EFFORT CONT.	TRACTIVE-EFFORT HOURLY	MPH CONT.	-MPH HOURLY	MAXIMUM SAFE MPH	CLASS
EF-1	576,000	451,000	80,800	230	285	3340	4100	80,000	106,000	15.5	14.5	45	EF-1
EF-2	864,000	676,000	122,000	230	285	5010	6150	121,200	159,000	15.5	14.5	45	EF-2
EF-3	816,000	691,000	124,500	230	285	5010	6150	121,200	159,000	15.5	14.5	45	EF-3
EF-4	586,800	435,400	78,300	345	375	5110	5530	77,000	85,500	25.2	24.5	65	EF-4
EF-5	1,056,000	931,000	168,000	230	285	6680	8200	161,600	212,000	15.5	14.5	45	EF-5
EP-1A	658,500	523,500	94,100	230	285	3340	4100	43,100	60,000	29.0	26.5	75	EP-1A
EP-2	521,200	457,800	82,200	222	247	3180	3517	42,000	48,500	28.4	27.1	70	EP-2
EP-3	620,000	420,000	75,700	315	390	3400	4200	49,000	66,000	26.0	23.8	70	EP-3
EP-4	578,350	432,100	77,800	345	375	5110	5530	77,000	85,000	25.2	24.5	70	EP-4

SOURCE: Conference Paper for the American Institute of Electrical Engineers, by Laurence Wylie, 1955.

HORSEPOWER

Horsepower is a measure which reflects a locomotive's ability to accelerate rapidly and to pull a train at speed. Because total pulling power is determined by tractive efffort as well as by horsepower, railroads seldom use horsepower as a basis of comparing locomotives.

Steam engines were seldom rated in horsepower and when they were, the basis of measurement often was not uniform. Sometimes boiler horsepower was measured, sometimes cylinder horsepower and sometimes drawbar horsepower. Drawbar horsepower is that horsepower which is actually available for pulling trains. Drawbar horsepower and rail horsepower are the same.

Diesels are generally rated in terms of prime mover horsepower. Their rail horsepower is, however, only 85% of that amount. 15% of the horsepower delivered by the prime mover is used up in the mechanical process of driving a generator and then converting electric power into wheel rotation.

Electrics have always been rated in terms of rail horsepower except for some recent American products. General Electric's E60C series used on Amtrak and the Black Mesa are rated in terms of "diesel equivalent horsepower." Thus they are advertised as 6,000 horsepower electrics although they only produce 5,100 at the rail.

Though horsepower is a useful statistic, the fallacy of trying to equate it with pulling power is most clearly illustrated when European locomotives are compared with those in America. Many of the high speed European locomotives are rated at 5,000-10,000 horsepower, but they lack tractive effort. The tonnage they can roll is quite limited, but they roll it very fast.

For train speeds below 20 mph, a locomotive's pulling power is determined almost entirely by tractive effort. A low powered switch engine will pull as much as a high powered road locomotive as long as their tractive effort is equal.

Horsepower requirements rise rapidly as train speed increases. Assuming a locomotive team has sufficient tractive effort to start and roll a 5,000 ton train on level track, approximately 6,000 rail horsepower would be required to reach 40 mph. To achieve 50 mph would require 7,500 horsepower. 60 mph would require 10,000 and 75 mph would require 15,000.

Initially assigned electric locomotive horsepower ratings are often based on design calculations made by the manufacturer. If in-service performance proves to be different than what was expected, the ratings may be revised. An example is the Little Joe. When these entered service in 1950, they were conservatively rated at 4,750 horsepower continuously and 5,120 for one hour. Experience showed that these locomotives could actually produce 5,110 horsepower continuously and 5,530 for one hour. Those figures were for performance at 3,000 volts. Higher or lower voltage resulted in higher or lower horsepower. The Little Joe rating at 3,300 volts was 5,550 horsepower continuously and 5,860 for one hour. Voltage along the line varied from 2,000 volts to 3,300 volts when measured at the locomotive. Low voltage was caused primarily by overloading of the railroad power supply system. Because of this type of voltage fluctuation, electric locomotive horsepower was somewhat variable in train service. It was not a fixed number.

TRACTIVE EFFORT

Tractive effort is a measure of how much pulling force a locomotive can exert. If the locomotive tends to slip its drivers, tractive effort will reflect how much traction the locomotive can get. If the locomotive drivers do not slip, tractive effort will reflect how much pull the locomotive can exert before stalling or overheating its traction motors.

Generally, a steel tired vehicle running on steel rails cannot exert a pull any greater than one quarter of the weight resting on its driving wheels. For that reason, tractive effort at 25% adhesion has become a standard locomotive statistic.

The Milwaukee Road conducted few formal tests of tractive effort in electrified territory, but the Great Northern conducted a number of them using steam, diesel and electric locomotives. The Milwaukee's historical experience concurred with the results of the Great Northern tests.

The Great Northern found that locomotives could typically achieve 25% to 40% adhesion when starting a train on good, dry rail. Such adhesion typically required that the wheels and rails have no grease film and that sand was continuously applied to increase the level of available traction. The Great Northern also found that regardless of claims made by locomotive manufacturers, no type of locomotive could regularly exceed 18% to 20% adhesion in a continuous over the road pull. That level of adhesion required good rail and good weather. The all weather adhesion was lower. For instance, four axle diesels averaged 12% to 14%, while six axle diesels averaged 14% to 16%, when weather was a variable factor.

When the Milwaukee tested the demonstrator Little Joe, GE-750, the highest tractive effort recorded was 35%. This was recorded under ideal test conditions on the 2.2% grade near Beverly, Washington, on the Coast Division. That level of adhesion with a Little Joe equals 152,390 pounds of tractive effort. Since the one hour rating for a Little Joe was 85,500 pounds, that pull could not have been sustained for very long without burning up the traction motors.

When tractive effort in excess of 20% is demanded, steam engines tend to slip because their uneven driving forces make the wheels break traction and spin. Most diesels and electrics have a smooth drive, but their independently driven axles can slip independently. Wheel slip tends to start with one pair of drivers and spread to others. One pair will momentarily lose its grip and spin. The other drivers then begin to slip because the weight of the train is too great to pull unless all drivers are pulling.

Wheel slip in a diesel or electric locomotive can damage the traction motors by allowing them to turn at excessive speeds. Centrifugal force may then tear the motor armatures apart. Wheel slip on any locomotive can cause train delays. A passenger locomotive cannot keep up with its schedule if wheel slip holds back acceleration. A freight locomotive may simply stall and tie up traffic on the line if it cannot get traction. Train dispatchers base their assignments of motivepower on dependable all weather tractive effort figures in order to avoid these problems.

SHORT-TIME HORSEPOWER AND TRACTIVE EFFORT RATINGS

The power produced by an electric locomotive is not limited by the capacity of any onboard steam or diesel-electric generating system. It is limited only by the energy available in the trolley and the capacity of its traction motors. The continuous rating for traction motors is the maximum amount of power they can deliver indefinitely without overheating. The short time ratings such as one hour, 30 minutes or 15 minutes do involve overheating. It takes time for a traction motor to get so hot that it burns up. How much time depends on the thermal capacity of the motor and how much extra power is demanded of it. A GE Motor could deliver 123% of its normal horsepower for one hour before its traction motors became dangerously hot. It could deliver approximately 171% of its normal horsepower for 30 minutes.

A train on level track can usually reach full speed in 10 or 15 minutes. A train could often crest a mountain grade on the Milwaukee within an hour. This underscores the value of short-time horsepower and tractive effort potential. A locomotive, like an automobile, is seldom operated flat out. If, however, a burst of power is needed an electric locomotive can provide it. The older electrics could deliver more short time power than a Little Joe because their traction motors were over designed. Diesel traction motors can handle so little overload because they are designed to work with diesel power plants which normally cannot generate enough current to overload them.

ONE HOUR OVERLOAD RATINGS

	HORSEPOWER	TRACTIVE EFFORT
GE MOTOR	123%	132%
BIPOLAR	110%	115%
WESTINGHOUSE	124%	135%
LITTLE JOE	108%	110%
F-7 DIESEL	85%	102%

NOTE: Diesel horsepower is normally measured at the drive shaft of the diesel prime mover. If measured at the rail where electrics are measured, it is only 85% of prime mover horsepower. Since the prime mover cannot exceed its rating, the continuous rating and short time rating for horsepower are the same. For example, a 1500 horsepower diesel produces 1275 horsepower at the rail.

LOCOMOTIVE TONNAGE RATINGS

The true comparative measure of locomotives is how much they can pull. Railroads list this in tons and call it a tonnage rating. These ratings tell what size train a locomotive can pull while making the speed the dispatcher wants to see. Train weight does not equate directly with train length, so pulling power is seldom expressed in numbers of cars. During the 1940's and 1950's, it was the general practice of Milwaukee Road dispatchers to estimate the weight of loaded freightcars at 50 tons. Empties were estimated at 22 tons. Heavy weight passenger cars averaged 80 tons each, while the post-World War II streamliners averaged 65 tons each.

Locomotive tonnage ratings are assigned between two points, usually the foot and summit of a grade. If the entire subdivision has a relatively constant gradient, then tonnage ratings may be listed between distant terminals. For through trains on routes with varying gradients the lowest tonnage rating for the route governs.

If the route has a gentle downgrade and tonnage is not limited by locomotive pulling and braking power, then the car limit applies. The car limit is determined by siding capacities. In order to pass each other along the line, trains must be able to fit into the sidings. Few sidings on the Rocky Mountain Division exceed 120 cars in length and on the Coast Division, few exceeded 110.

COAST DIVISION
TONNAGE RATINGS TACOMA TO OTHELLO

	RULING GRADE	EP-2 Bipolar	EF-1 2 unit GE Motor	EF-2/EF-3 3 unit GE Motor	EF-5 4 unit GE Motor	Diesel 2 unit FT	Diesel 4 unit FT	S-1 4-8-4	N-2 2-6-6-2	N-3 2-6-6-2	L-2 2-8-2	F-5 4-6-2	C-2/C3 C-5 2-8-0
EASTBOUND													
Tacoma-Black River	0.0%	3450	CL	CL	CL	CL	CL	4500	CL	CL	4000	3000	3000
Black River-Cedar Falls	0.8%	2000	4100	5500	8200	2750	5500	2300	2400	2750	2000	1500	1500
Cedar Falls-Hyak	1.74%	1250	1550	2550	3400	1325	2650	975	1150	1300	950	700	700
Hyak-Cle Elum	down	CL	CL	CL	CL	CL	CL	CL	CL	CL	CL	CL	CL
Cle Elum-Kittitas	down	4000	6000	7500	12000	CL	CL	4500	5000	6000	4500	3500	3500
Kittitas-Boylston	1.6%	1300	1670	2500	3340	1450	2900	1000	1200	1360	960	740	840
Boylston-Beverly	down	1300	1670	2500	3340	925	1850	CL	CL	CL	CL	CL	CL
Beverly-Othello	0.4%	3200	5000	7000	10000	3500	CL	3700	3900	4300	3000	2600	2600
WESTBOUND													
Othello-Beverly	down	CL	CL	CL	CL	CL	CL	CL	CL	CL	CL	CL	CL
Beverly-Boylston	2.2%	980	1200	1800	2800	1025	2050	700	900	1000	700	550	550
Boylston-Kittitas	down	1400	3100	4650	6200	1425	2850	CL	CL	CL	CL	CL	CL
Kittitas-Cle Elum	0.4%	3700	5000	7000	10000	4200	CL	3100	3300	3700	2600	2000	2000
Cle Elum-Hyak	0.7%	3200	4000	5500	8000	3075	6150	2600	2700	3100	2250	1700	1700
Hyak-Cedar Falls	down	1250	2800	4000	5600	1350	2700	CL	CL	CL	CL	CL	CL
Cedar Falls-Black River	down	CL	CL	CL	CL	CL	CL	CL	CL	CL	CL	CL	CL
Black River-Tacoma	0.0%	3450	CL	CL	CL	CL	CL	4500	CL	CL	4000	3000	3000

STEAM ENGINE TONNAGE REDUCTIONS FOR COLD WEATHER

10-20 Above .... Reduce 10%
Zero-10 Above .... Reduce 15%
Zero-10 Below .... Reduce 20%
10-20 Below .... Reduce 30%
CL = Car Limit
EF-5 According to 1959 Employee Timetable, all others according to 1947 Employee Timetable

IDAHO DIVISION
TONNAGE RATINGS OTHELLO to AVERY (non-electrified)

	RULING GRADE	DIESEL 4-unit FT	N-3 2-6-6-2	S-1 4-8-4	F-6 4-6-4	F-5 4-6-2	C-3 2-8-0
EASTBOUND via MALDEN							
Othello-Avery	0.4%	6800	5543	4000	3000	2706	3018
WESTBOUND via MALDEN							
Avery-Ramsdell	1.0%	4600	2815	1920	1675	1462	1539
Ramsdell-Sorrento							
Sorrento-Marengo Jct							
Marengo Jct-Hillcrest	0.7%	8400	3771	2700	2050	1832	2057
Hillcrest-Othello							
EASTBOUND via SPOKANE							
Marengo Jct-Spokane							
Spokane—Chester							
Chester-Manito	1.7%	3000	1700	1360	1110	813	933
Manito-Plummer Jct	0.75%	5500	3574	2500	1935	1735	1950
WESTBOUND via SPOKANE							
Plummer Jct-Worley	1.0%	4600	2815	1920	1700	1462	1539
Worley-Spokane	0.7%	5600	3771	2700	2050	1832	2057
Spokane-Cheney	0.69%	6000	2860	2750	2100	1853	2079
Cheney-Marengo Jct							

STEAM ENGINE TONNAGE REDUCTIONS FOR COLD WEATHER

10-20 Above.................... Reduce 10%
Zero-10 Above .... Reduce 15%
Zero-10 Below .... Reduce 20%
10-20 Below .... Reduce 30%

(Ratings according to 1947 Employees Timetable)

ROCKY MOUNTAIN DIVISION
TONNAGE RATINGS AVERY TO HARLOWTON

	RULING GRADE	EP-1A Rebuilt GE Passenger Motor	EP-2 Rebuilt Bipolar	EP-3 Westing-house Motor	EP-4/ EF-4 Little Joe	EF-1 2 unit GE Motor	EF-2/ EF-3 3 unit GE Motor	EF-5 4 unit GE Motor
EASTBOUND								
Avery-East Portal	1.7%	1250	1250	1150	1600	1750	2625	3500
East Portal-St. Regis	down							
St. Regis-Deer Lodge	0.4%			3500	5400	6000	9000	12000
Deer Lodge-Alloy	0.6%			3000	4050	4500	6750	9000
Alloy-Donald	1.66%	1400	1400	1150	1600	1750	2625	3500
Donald-Lombard	down							
Lombard-Cardinal	0.46%			3500	5400	6000	9000	12000
Cardinal-Loweth	1.0%	1600	1600	1600	2600	2650	3975	5300
Loweth-Harlowton	down							
WESTBOUND								
Harlowton-Valencia	0.6%			3000	4050	4500	6750	9000
Valencia-Bruno	1.0%			1600	2520	2800	4200	5600
Bruno-Loweth	1.4%	1300	1300	960	2250	2400	3600	4800
Loweth-Lombard	down							
Lombard-Piedmont	0.3%			4000	7200	8000	12000	16000
Piedmont-Donald	2.0%	1050	1050	960	1600	1500	2250	3000
Donald-St. Regis	down							
St. Regis-Haugan	0.8%			1600	2520	2800	4200	5600
Haugan-Roland	1.7%	1250	1250	1150	1700	1750	2625	3500

GE Freight Motors limited to maximum of 45 mph
EP-1, EP-2, EP-3, EP-4/EF-4— According to 1956 Employee Timetable
EF-1, EF-2, EF-3, EF-4, EF-5— According to 1967 Employee Timetable

ROCKY MOUNTAIN DIVISION

TONNAGE RATINGS AVERY TO HARLOWTON

	RULING GRADE	GP9 1750 hp	GP35 2500 hp	U25B 2500 hp	GP40 3000 hp	U30B 3000 hp	SD45 3600 hp	U36C 3600 hp	Little Joe 6000 hp*
EASTBOUND									
Avery-East Portal	1.7%	665	990	1050	1080	1270	1460	1480	1600
East Portal-St. Regis	down								
St. Regis-Deer Lodge	0.4%	2330	3340	3510	3630	4200	4930	4970	5400
Deer Lodge-Alloy	0.6%	1730	2490	2630	2720	3150	3680	3710	4050
Alloy-Donald	1.66%	710	1050	1110	1150	1350	1550	1570	1600
Donald-Lombard	down								
Lombard-Cardinal	0.46%	1990	2860	3010	3110	3600	4220	4250	5400
Cardinal-Loweth	1.0%	1110	1630	1730	1780	2070	2410	2430	2600
Loweth-Harlowton	down								
WESTBOUND									
Harlowton-Valencia	0.6%	1730	2490	2630	2720	3150	3680	3710	4050
Valencia-Bruno	1.0%	1110	1630	1730	1780	2070	2410	2430	2520
Bruno-Loweth	1.4%	815	1200	1270	1310	1530	1770	1790	2250
Loweth-Lombard	down								
Lombard-Piedmont	0.3%	2800	4000	4220	4350	5030	5910	5960	7200
Piedmont-Donald	2.0%	560	835	890	920	1080	1240	1260	1600
Donald-St. Regis	down								
St. Regis-Haugan	0.8%	1350	1980	2090	2160	2510	2920	2950	2520
Haugan-Roland	1.7%	665	990	1050	1080	1270	1460	1480	1700

*Horsepower listed for Little Joe is "diesel equivalent horsepower."

Diesel horsepower measured at the rail is only 85% of that produced by the prime mover. A Little Joe operating at 3,000 volts produces 5,100 horsepower and 5,100 is 85% of 6,000. Diesel speed at rated tonnage is 18 mph, Little Joe speed at rated tonnage is 24 to 28 mph.

SOURCE OF DIESEL TONNAGE RATINGS: Milwaukee Road Load-Speed Rating Charts

OBSERVATIONS ON LOCOMOTIVE NUMBERING

The Milwaukee's electric locomotive numbering system was as straight forward as any in labeling the single unit locomotives. The Bipolars, Westinghouse Motors, and Switchers were only renumbered once. New numbers were assigned within each class sequentially except for the Westinghouse Motors. The Little Joes always bore the same numbers.

When the numbering system was applied to the multiple unit GE Motors, it became confusing. It was not confusing to railroaders, only to historians and railfans. The system was designed to label each locomotive but it was not designed to provide for historical tracking of units assigned to each locomotive. The Milwaukee considered it important that no two locomotives bear identical numbers at the same time. However, different locomotives could bear identical numbers at different times or change numbers without following a master plan. For railroaders, the important factor was that on a given date all reports mentioning a particular locomotive were discussing the same locomotive. Because of renumberings and the movement of units from locomotive to locomotive, an all-time roster of units is complex.

The source of the numbering complexities arose when the GE Motors were first constructed. The 84 units were built as 42 locomotives, each having two halves. The Milwaukee designated each half a "half unit" and gave each half unit the same road number followed by the letter A or B.

The Passenger Motors were 10100 A & B through 10111 A & B. The Freight Motors were 10200 A & B through 10229 A & B. In 1920, after the arrival of the Bipolars and the Westinghouse Motors, the GE Passenger Motors were regeared for freight service and became 10230 A & B through 10241 A & B.

In 1932, the Milwaukee began creating three-unit locomotives, an activity that continued until 1954. The Milwaukee did this by placing a third "half-unit" between the two half-units of an A and B set. The three-unit locomotives were assigned numbers 10500 ACB through 10511 ACB.

From 1936 through 1939, carbodies were shortened and pilot trucks were removed from twelve middle units. A chain of renumberings occurred as units were rotated through the shops one by one. Shortened units were installed in the sets as standard units went into the shops.

In March of 1939, the Milwaukee renumbered its locomotive fleet and the GE Freight Motors became E25 through E73. There was no longer a separate series for three unit motors. From 1940 until 1954, the Milwaukee continued to create three unit sets, however no more middle units had their carbodies shortened and pilot trucks removed.

In the ensuing years, various sets and individual units were renumbered, some, several times. From 1951 through 1961, the primary causes of GE Motor renumberings were the creation of 13 four-unit freight locomotives, a two-unit passenger locomotive and a three-unit passenger locomotive. This required reshuffling the fleet. Four-unit sets were assigned the road number of the A and B units and the sets were generally arranged ACDB. A's and B's remained as control units.

In the late 1960's, age and wear began to take its toll. The total number of GE Motors began to shrink as worn out units were scrapped. While scrapping progressed the remaining units were shifted around to keep locomotives running as three and four unit sets. Sometimes the shifted units were renumbered to match the set they ran with. Other times they retained their old numbers, resulting in locomotives containing odd numbered units. During the early

1970's units continued to wear out. Locomotives shrank from four units to three units to two units. By 1974, when the end came only one set was running, the two unit set of E57B and E34C, which was running as the Harlowton Switcher. Two years earlier E34C had been known as E47D.

When reviewing rosters and locomotive assignments, there is another important fact to keep in mind. The Milwaukee's operating practices were dictated by business necessity rather than strict adherance to fixed policies. Thus Coast Division electrics might be temporarily assigned to the Rocky Mountain Division to handle extra traffic. Rocky Mountain Division electrics might be seen on the Coast Division while traveling to the Tacoma shops. Coast Division mallets might be assigned to the Idaho Division to cover for an Idaho Division mallet which was undergoing repairs. When GE Motors came due for shopping, not all units required work at the same time. Thus a GE Motor which normally ran as a three or four unit set might temporarily run as a two or three unit set.

MAINLINE ELECTRIC ROSTER

TYPE OF LOCOMOTIVE	NUMBER OF UNITS OWNED	ROAD NUMBERS PRIOR TO 1939	ROAD NUMBERS AFTER 1939
GE Motors	84	Passenger Motors = 10100-10111 (converted to freight = 10230-10241) 2-unit Freight Motors 10200-10241 3-unit Freight Motors 10500-10511	Passenger E22 and E23 Freight E22-E73
Switchers	4	10050-10053	E80-E83
Bipolars	5	10250-10254	E1-E5
Westinghouse Motors	10	10300-10309	10300-10302 = E10-E12 10303 = E19 10304-10309 = E13-E18
Little Joes	12	(Purchased 1950)	Passenger E20 and E21 Freight E70-E79 All twelve used in freight service after 1958

ROSTER OF GE MOTORS

EF-5 FOUR UNIT FREIGHT MOTORS 1950-1961

ROAD NUMBER	ARRANGE-MENT	DIVISION ASSIGNED	BECAME 4 UNITS	CEASED TO BE 4 UNITS	COMMENTS
E22	ABCD	Coast	1961		
E25	ADCB	Coast	1951		
E30	ACDB	Coast	1953		
E32	DACB	Coast	1952	1961	Became 3 units E32C renumbered E47D
E33	ACDB	Coast	1960		
E34	ACDB	Rocky	1951		
E36	ACDB	Rocky	1960		
E37	ACDB	Rocky	1951	1960	Became 2 units, E37 C & D renumbered E36 C & D
E39	ADCB	Coast	1960		
E45	ACDB	Rocky	1951		
E47	ACDB	Coast	1961		
E49	ACDB	Rocky	1951		
E50	ADCB	Coast	1960		

EF-3 THREE UNIT FREIGHT MOTORS 1950-1961

ROAD NUMBER	ARRANGE-MENT	DIVISION ASSIGNED	ASSIGNED THIS NUMBER	CEASED TO BE 3 UNITS	COMMENTS
E25	ACB	Coast	1939	1951	Became a 4 unit motor
E26	ACB	Rocky	1939	1951	Renumbered, E26A = E49D, E26B = E45D, E26C = E34D
E27	ACB	Rocky	1939		
E28	ACB	Coast	1939	1951	Became 2 unit motor, E28C = E25D
E29	ACB	Rocky	1939		
E30	ACB	Coast	1939	1953	Became a 4 unit motor
E31	ACB	Coast*	1939	1952	Became a 2 unit motor, E31C = E33D
E32	ACB	Coast	1939	1952	Became a 4 unit motor
E33	ACB	Coast	1939	1952	Became a 4 unit motor
E34	ACB	Rocky	1939	1951	Became a 4 unit motor
E35	ACB	Rocky	1943		
E36	ACB	Rocky	1939	1953	Became EF-2, E36C = E30D, E56A = A36C

EF-2 THREE UNIT FREIGHT MOTOR 1950-1961

E31	ACB	Coast	1954		
E32	DAB	Coast	1961		E32C Renumbered E47D
E36	ACB	Rocky	1953	1955	Became 2 unit motor, E36C = E23C
E37	ACB	Rocky	1940	1951	Became a 4 unit motor
E38	ACB	Rocky	1943		
E39	ACB	Coast	1946	1960	Became a 4 unit motor
E40	ACB	Coast	1946		
E41	ACB	Coast*	1946	1954	E41C & B wrecked at Doris, E41A renumbered E31C
E42	ACB	Coast	1950		
E45	ACB	Rocky	1942	1951	Became a 4 unit motor
E47	ACB	Coast*	1942	1961	Became a 4 unit motor
E49	ACB	Rocky	1943	1951	Became a 4 unit motor
E52	ACB	Rocky	1939		
E54	ACB	Rocky	1952		

*These motors previously ran on the Rocky Mountain Division and were reassigned to the Coast.

EF-1 TWO UNIT FREIGHT MOTORS 1950-1961

ROAD NUMBER	ARRANGE-MENT	DIVISION ASSIGNED	ASSIGNED THIS NUMBER	CEASED TO BE 2 UNITS	COMMENTS
E28	AB	Coast	1951[1]	1953	Became E23 AB
E31	AB	Rocky	1952	1954	Became 3 units motor
E36	AB	Rocky	1955	1960	Became 4 unit motor
E37	AB	Rocky	1940	1964	Scrapped
E50	AB	Coast	1939	1960	Became 4 unit motor
E51	AB	Coast	1939	1951	Wrecked in Saddle Mountains
E55	AB	Rocky	1939	1951	E55A, E55B = E37D became Harlowton Switcher
E56	AB	Rocky	1939	1953	E56A renumbered E36C, E56B Harlowton Switcher
E57	AB	Rocky	1939	1950	E57A renumbered E42C, E57B Harlowton Switcher
E59	AB	Rocky	1939	1952	Both units renumbered E59A = E52C, E59B = E54C
E64	AB	Coast	1939	1952	Both units renumbered E64A = E32D, E64B = E50D
E68	AB	Coast	1943	1951	Wrecked at Auburn
E69	AB	Coast	1950	1953	Renumbered E22 AB

FREIGHT MOTOR UNITS WITH SHORTENED CARBODY 1950-1961

E25C, E25D, E26C, E27C, E28C, E29C, E30C, E30D, E31C, E32C, E33C, E33D, E34C, E34D, E35C, E36C, E47D
(Twelve units total)

NOTE: Due to renumberings, E28C = E25D, E31C = E33D, E26C = E34D, E32C = E47D, E36C = E30D.

Four different units bore the number E36C. Three of them carried that number between 1950 and 1961. Only one of these units had a shortened carbody.

ES-3, SINGLE UNIT GE FREIGHT MOTORS 1950-1961
(also known as EF-1 half-units)

ROAD NUMBER	UNIT	SINGLE UNIT	OUT OF SERVICE	COMMENTS
E55	A	1951	1964	Harlowton Switcher, Scrapped
E56	B	1953	1960	Harlowton Switcher, Renumbered E50C
E57	B	1950		Harlowton Switcher

EP-1A TWO UNIT GE PASSENGER MOTORS 1950-1961

ROAD NUMBER	ARRANGE-MENT	ASSIGNED THIS NUMBER	RENUM-BERED	COMMENTS
E22	AB	1953	1960	Converted to freight
E23	AB	1953	1955	Unit C added 1955

EP-1A THREE UNIT GE PASSENGER MOTORS 1950-1961

ROAD NUMBER	ARRANGE-MENT	BECAME 3 UNITS	RENUM-BERED	COMMENTS
E23	ABC & ACB	1955	1960	Converted to freight, E23A = E22D, E23B = E22C, E23C = E39D

Substation buildings on the Milwaukee Road were large electrical vaults. They had concrete roofs, substantial brick walls, and thick concrete floors. Beneath the floors were basements designed to provide air circulation to the electrical equipment. Most buildings had flat roofs and were constructed from one of two standard plans. The buildings provided for either two or three motor-generator sets, with transformers. Most design variations from building to building involved the location of doors and offices.

Five of the twenty-two buildings were substantially different from the rest and somewhat different from each other. These were the gabled roof buildings located in deep snow zones. They were located at Drexel, East Portal, Avery, Cle Elum and Hyak. East Portal was the largest of all substation buildings. Hyak was the tallest, resting atop a ten foot tall foundation. Drexel and Cle Elum had only one arched window in front rather than two. The diagram of Avery is included here because it is somewhat representative of the gabled roof buildings. The following is a list of buildings by floor plan type.

TWO M-G SET PLAN		THREE M-G SET PLAN
Cle Elum*	Primrose	Avery
Drexel	Ravenna	Cedar Falls*
Eustis	Renton	Doris
Francis	Tacoma Jct.*	East Portal
Gold Creek	Tarkio	Hyak*
Loweth	Taunton	Janney
Morel	Two Dot	Kittitas*
		Piedmont

*These buildings contained one less M-G set and transformer than they were designed to hold.

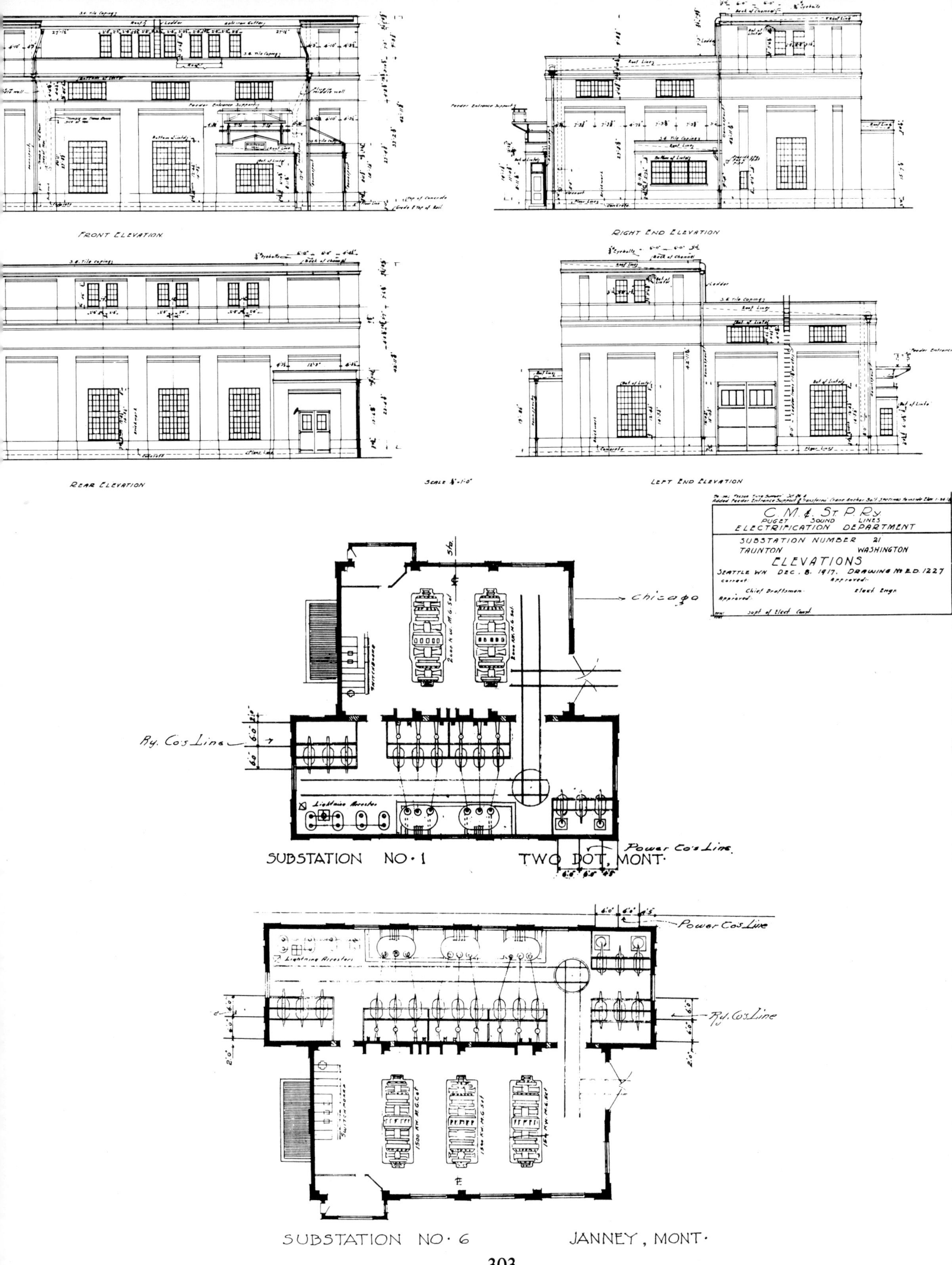
FRONT ELEVATION
RIGHT END ELEVATION
REAR ELEVATION
SCALE ⅛"=1'-0"
LEFT END ELEVATION
C.M. & St. P. Ry.
PUGET SOUND LINES
ELECTRIFICATION DEPARTMENT
SUBSTATION NUMBER 21
TAUNTON WASHINGTON
ELEVATIONS
SEATTLE WN. DEC. 8. 1917. DRAWING No. E.D. 1227
Chicago
Ry. Co's Line
Power Co's Line
Lightning Arrester
SUBSTATION NO·1 TWO DOT, MONT·
Power Co's Line
Lightning Arresters
Ry. Co's Line
Switchboard
1500 KW. M.G. Set
SUBSTATION NO·6 JANNEY, MONT·

ARRANGEMENTS OF TYPICAL MILWAUKEE ROAD TROLLEY SUPPORTS

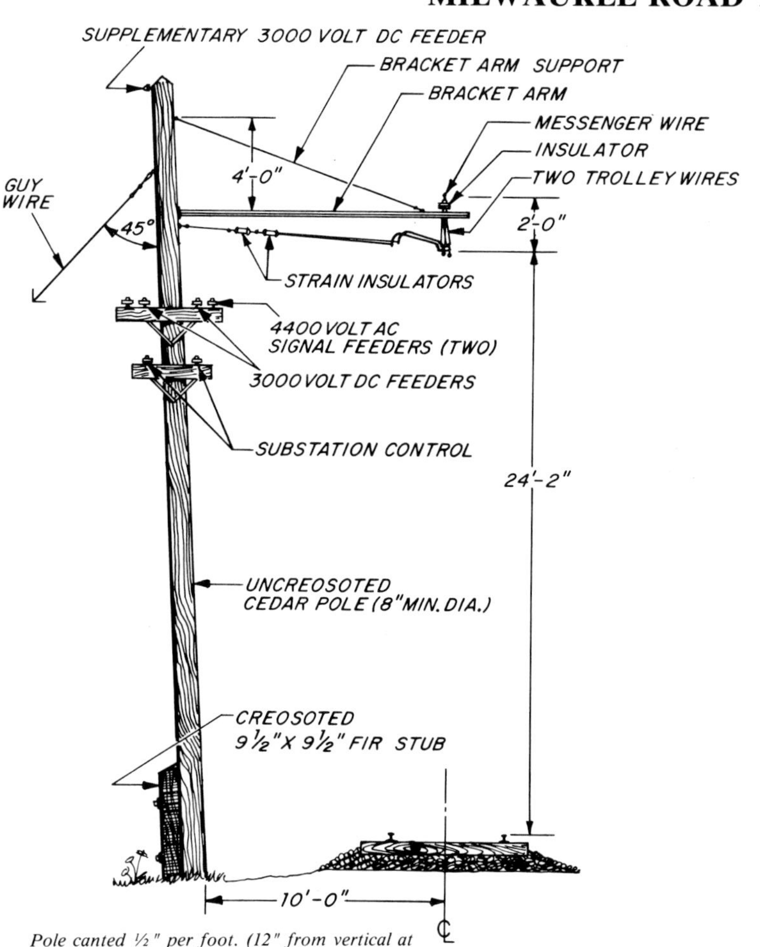

Pole canted ½" per foot. (12" from vertical at trolley wire height.)

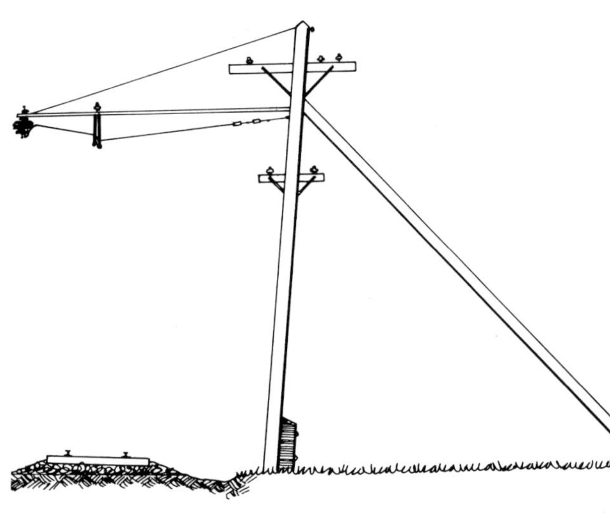

As a general rule, poles are not placed on the inside of a curve. However, if terrain necessitates placing them there, the pole shown above is used. This type of pole is braced rather than guyed and the pull-off is still arranged to pull toward the outside of the curve.

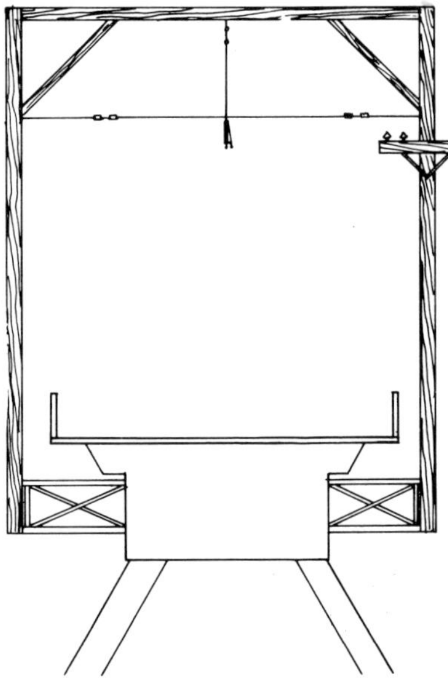

On trestles, the Milwaukee often supported the trolley with wooden portal structures. On some trestles, the Milwaukee used single poles attached to one side of the trestle.

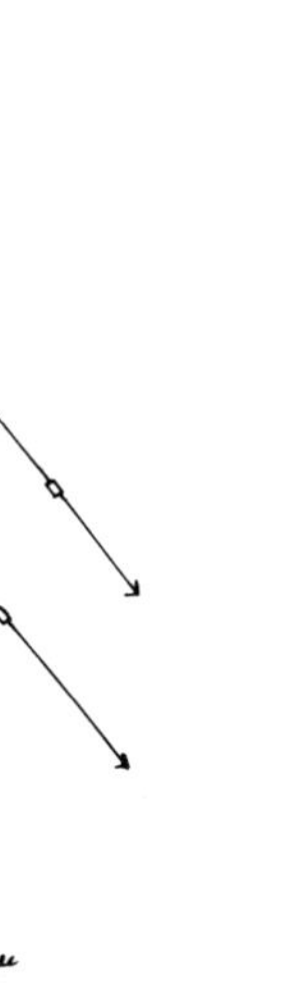

Span wires such as those shown above are generally used to support catenary above two or more tracks. The maximum number of tracks spanned in this manner was nine. Sometimes when several tracks were being spanned, a middle pole was also used. Catenary was not necessarily above all tracks spanned in a yard.

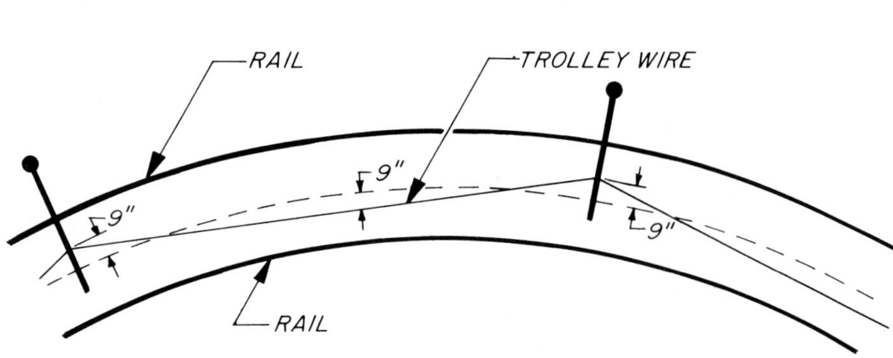

In tunnels, the Milwaukee hung the trolley from insulated metal brackets like the one shown above.

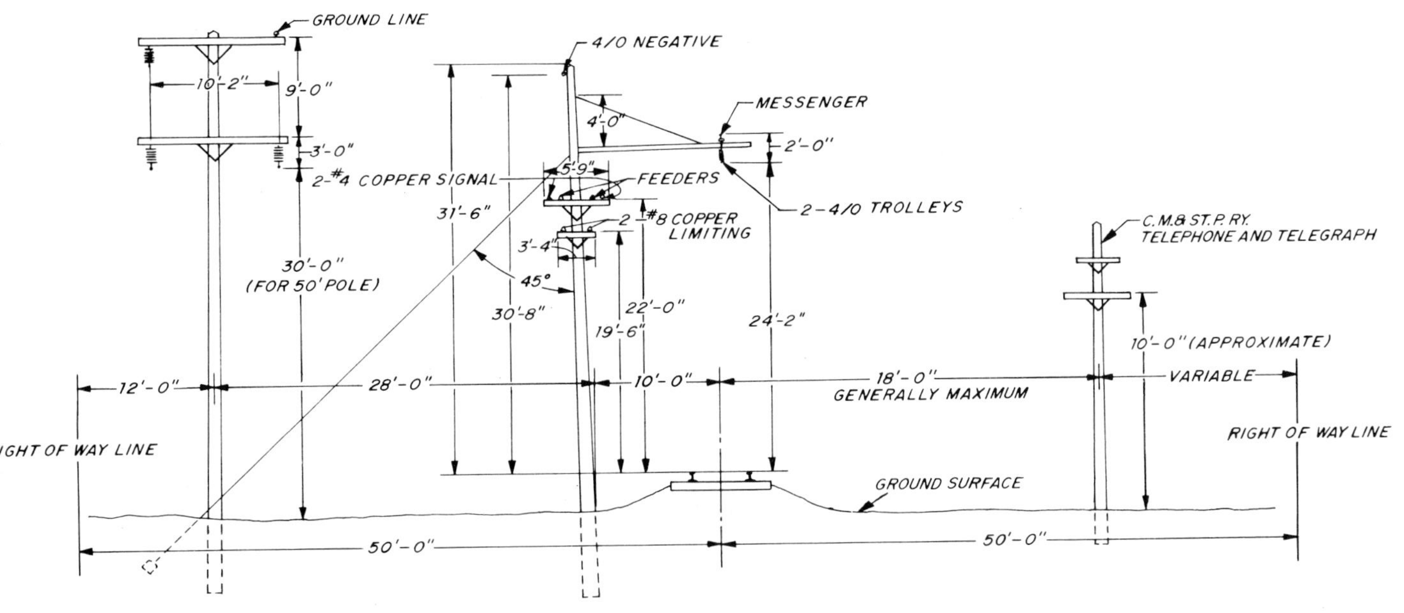

Transmission & Distribution System. Conventional Cross Section of Right of Way with 100' width.

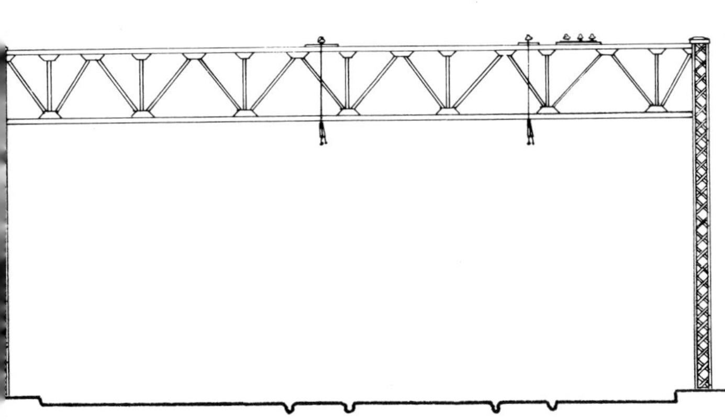

Steel bridge type trolley supports were found on the Milwaukee only in Renton, Washington, where the tracks ran down the center of Houser Street. The spacing of these supports was 350' rather than the 150' used elsewhere with wooden poles. At the time these bridges were built the street contained two tracks. One was later removed.

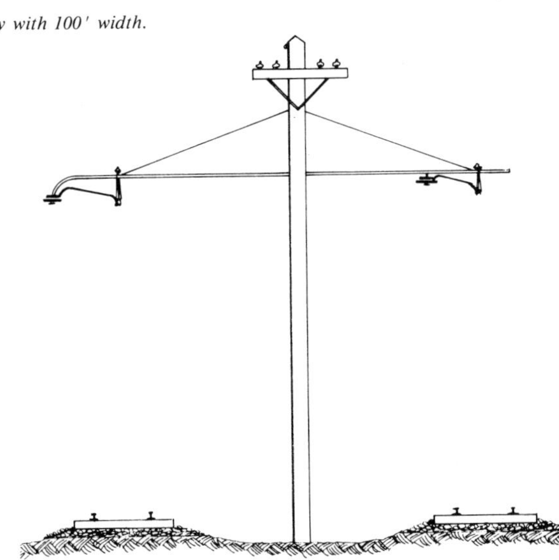

This type of double track trolley support was seldom used on the Milwaukee. It was the primary type of trolley support only along the ten miles of track between Seattle and Black River Junction.

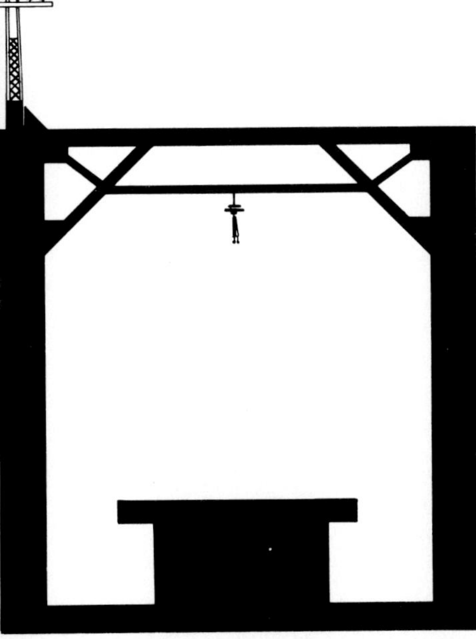

On truss bridges, the Milwaukee usually supported the trolley with a disc insulator hung from a steel beam.

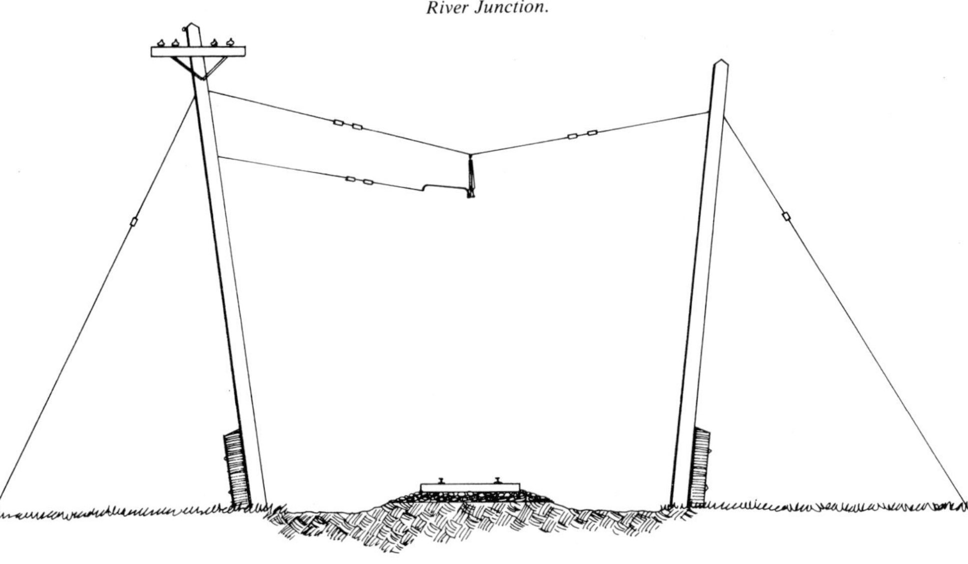

This type of span wire construction was used in single track areas where the trolley must follow "S" curves in the track. The one wire span was sometimes also used in double track areas.

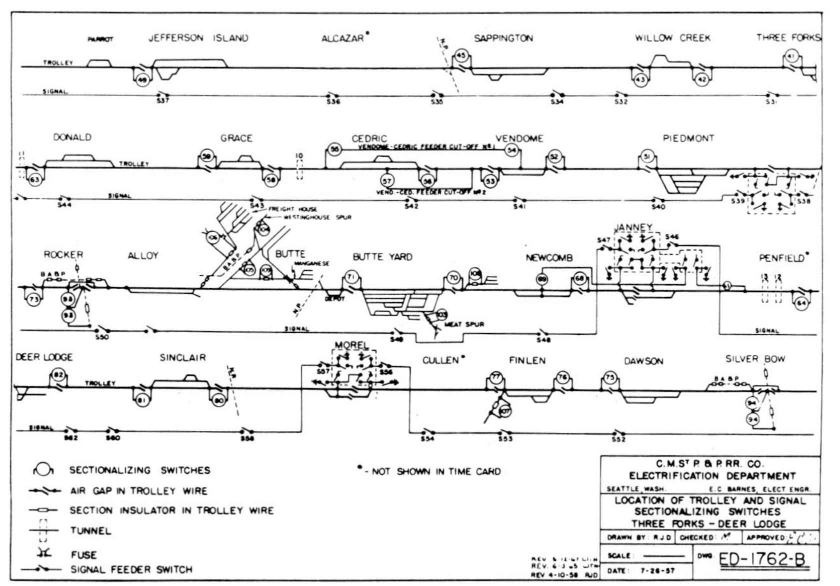
JEFFERSON ISLAND
ALCAZAR*
SAPPINGTON
WILLOW CREEK
THREE FORKS
DONALD
GRACE
CEDRIC
VENDOME
PIEDMONT
ROCKER
ALLOY
BUTTE
BUTTE YARD
NEWCOMB
JANNEY
PENFIELD*
DEER LODGE
SINCLAIR
MOREL
CULLEN*
FINLEN
DAWSON
SILVER BOW
SECTIONALIZING SWITCHES
AIR GAP IN TROLLEY WIRE
SECTION INSULATOR IN TROLLEY WIRE
TUNNEL
FUSE
SIGNAL FEEDER SWITCH
* - NOT SHOWN IN TIME CARD
C.M.St P. & P.RR. CO.
ELECTRIFICATION DEPARTMENT
SEATTLE, WASH.
E.C. BARNES, ELECT. ENGR.
LOCATION OF TROLLEY AND SIGNAL
SECTIONALIZING SWITCHES
THREE FORKS - DEER LODGE
DATE: 7-26-57
ED-1762-B

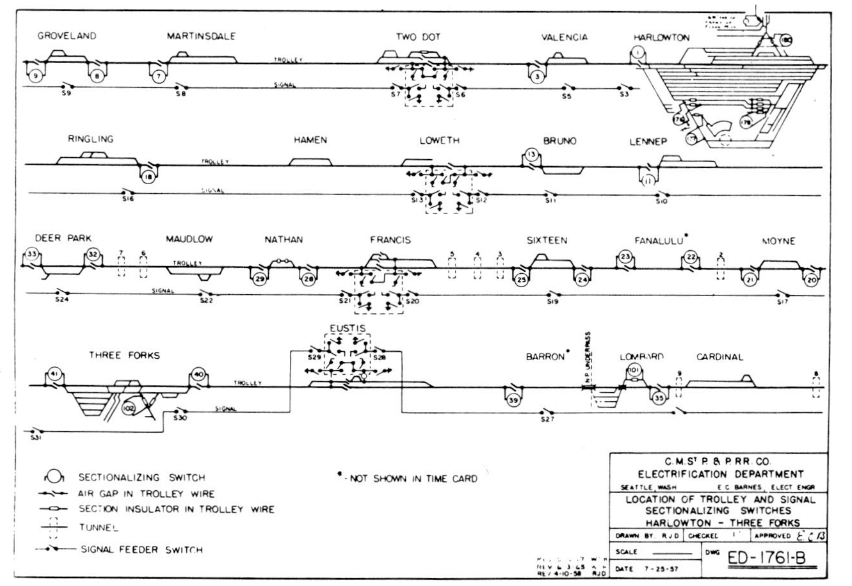
GROVELAND
MARTINSDALE
TWO DOT
VALENCIA
HARLOWTON
RINGLING
HAMEN
LOWETH
BRUNO
LENNEP
DEER PARK
MAUDLOW
NATHAN
FRANCIS
SIXTEEN
FANALULU*
MOYNE
THREE FORKS
EUSTIS
BARRON*
LOMBARD
CARDINAL
SECTIONALIZING SWITCH
AIR GAP IN TROLLEY WIRE
SECTION INSULATOR IN TROLLEY WIRE
TUNNEL
SIGNAL FEEDER SWITCH
* - NOT SHOWN IN TIME CARD
C.M.St P. & P.RR. CO.
ELECTRIFICATION DEPARTMENT
SEATTLE, WASH.
E.C. BARNES, ELECT. ENGR.
LOCATION OF TROLLEY AND SIGNAL
SECTIONALIZING SWITCHES
HARLOWTON - THREE FORKS
DATE 7-25-57
ED-1761-B

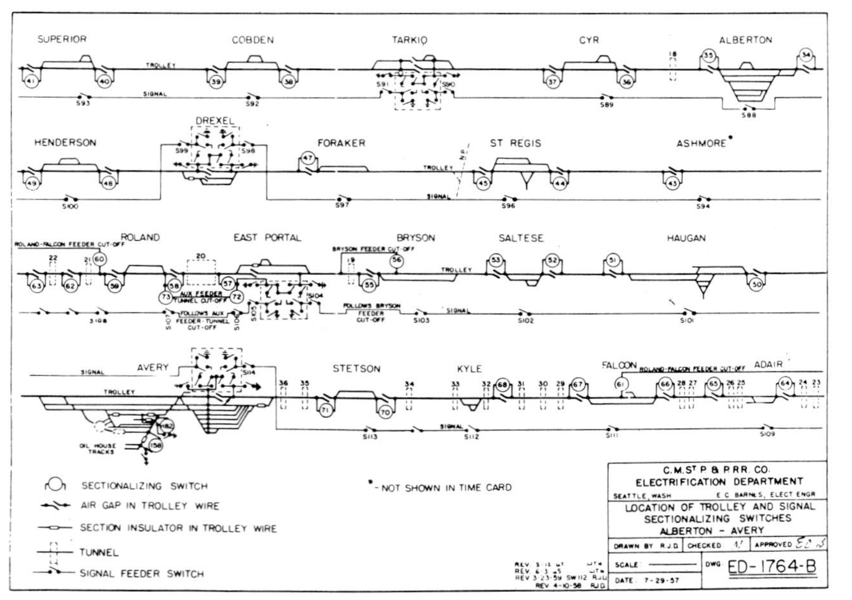
SUPERIOR
COBDEN
TARKIO
CYR
ALBERTON
HENDERSON
DREXEL
FORAKER
ST REGIS
ASHMORE*
ROLAND
EAST PORTAL
BRYSON
SALTESE
HAUGAN
AVERY
STETSON
KYLE
FALCON
ADAIR
SECTIONALIZING SWITCH
AIR GAP IN TROLLEY WIRE
SECTION INSULATOR IN TROLLEY WIRE
TUNNEL
SIGNAL FEEDER SWITCH
* - NOT SHOWN IN TIME CARD
C.M.St P. & P.RR. CO.
ELECTRIFICATION DEPARTMENT
SEATTLE, WASH.
E.C. BARNES, ELECT. ENGR.
LOCATION OF TROLLEY AND SIGNAL
SECTIONALIZING SWITCHES
ALBERTON - AVERY
DATE: 7-29-57
ED-1764-B

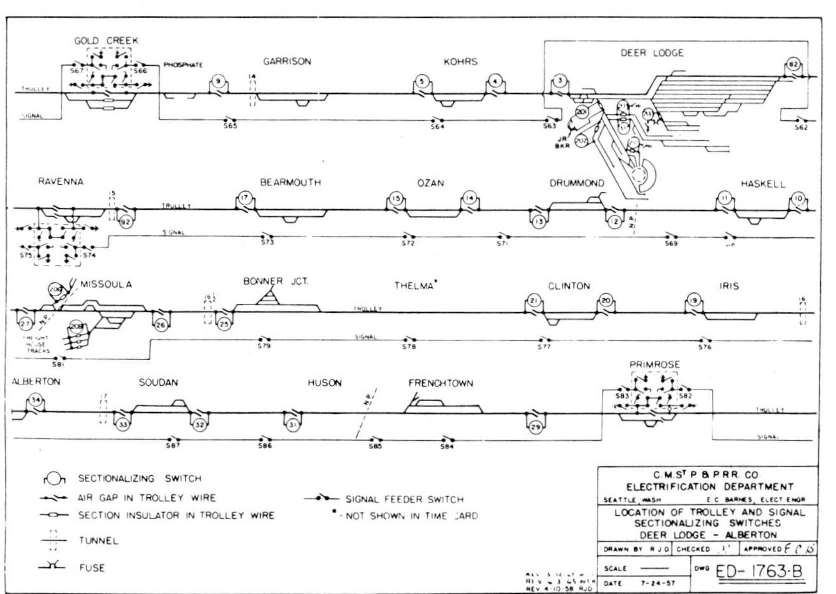
GOLD CREEK
GARRISON
KOHRS
DEER LODGE
RAVENNA
BEARMOUTH
OZAN
DRUMMOND
HASKELL
MISSOULA
BONNER JCT.
THELMA*
CLINTON
IRIS
ALBERTON
SOUDAN
HUSON
FRENCHTOWN
PRIMROSE
SECTIONALIZING SWITCH
AIR GAP IN TROLLEY WIRE
SECTION INSULATOR IN TROLLEY WIRE
TUNNEL
FUSE
SIGNAL FEEDER SWITCH
* - NOT SHOWN IN TIME CARD
C.M.St P. & P.RR. CO.
ELECTRIFICATION DEPARTMENT
SEATTLE, WASH.
E.C. BARNES, ELECT. ENGR.
LOCATION OF TROLLEY AND SIGNAL
SECTIONALIZING SWITCHES
DEER LODGE - ALBERTON
DATE 7-24-57
ED-1763-B

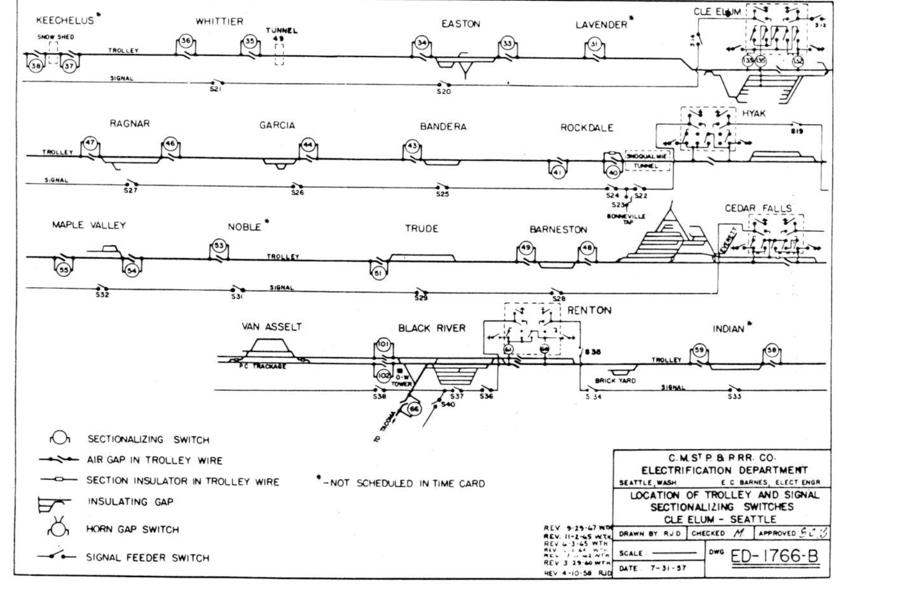

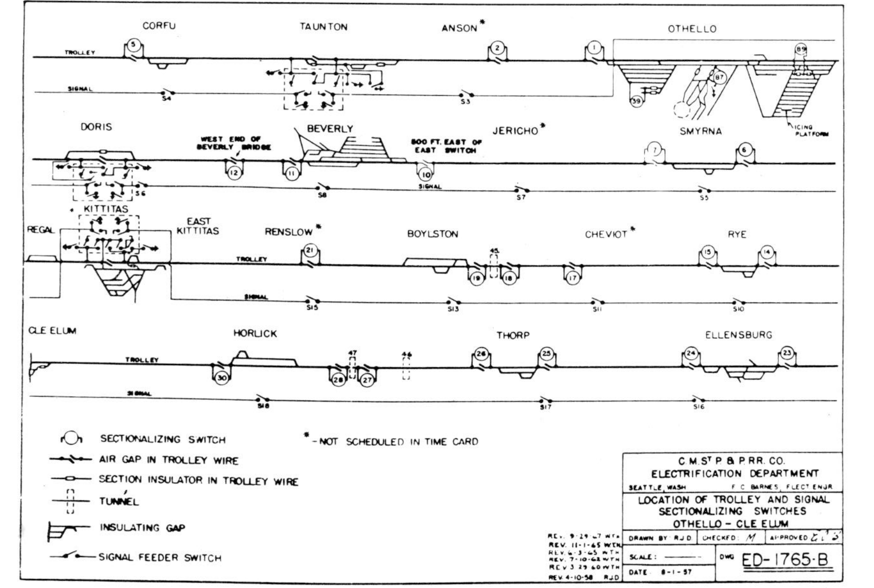

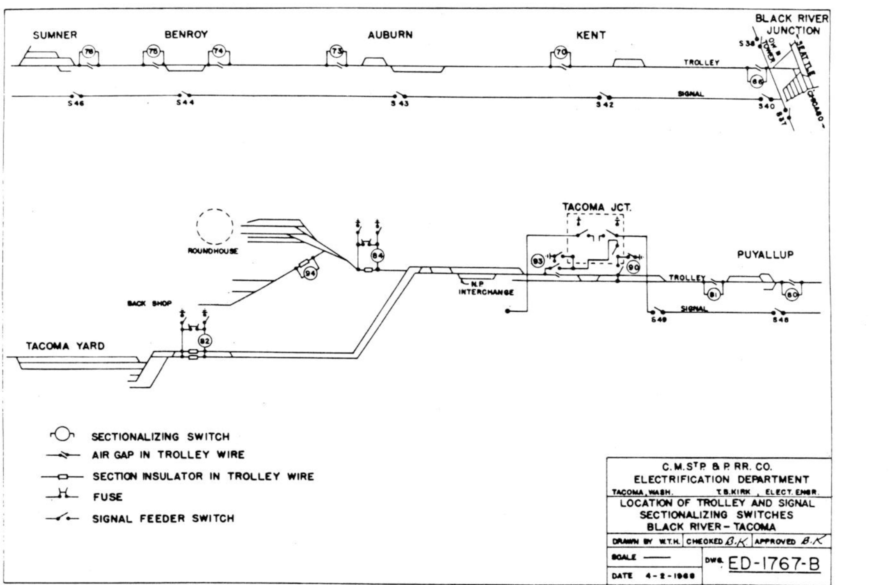

SCHEMATIC DIAGRAMS
SHOWING ALL ELECTRIFIED
TRACKAGE ON THE
MILWAUKEE ROAD
DRAWN 1957
LAST REVISED 1967

BIBLIOGRAPHY

BOOKS:

Brain, Insley J. *The Milwaukee Road Electrification.* The Western Railroader, 1961.

Derleth, August. *The Milwaukee Road: Its First Hundred Years.* New York: Creative Age Press, 1948.

Digest, Reader's. *Reader's Digest 1981 Almanac and Yearbook.* Reader's Digest Assn., 1981.

Electrification of Steam Railroads Committee. *Electrification of Steam Railroads.* National Electric Light Assn., 1930.

General Electric and Westinghouse. *Denver and Rio Grande Western Railroad Electrification. Denver and Pueblo to Salt Lake City.* General Electric Co. and Westinghouse Corp., March 3, 1947.

Manson, Arthur J. *Railroad Electrification and the Electric Locomotive.* Simmons-Boardman, 1925.

Mechanical Engineer's Office. *Locomotive Diagrams.* The Milwaukee Road, Revised 1960.

Middleton, William D. *South Shore: The Last Interurban.* Golden West Books, 1970.

The Railroad & Locomotive Historical Society, Inc. *Railroad History 136* (Milwaukee History and Roster) R&L HS, 1977.

Scribbins, Jim. *The Hiawatha Story.* Kalmbach Publishing Co., 1970.

Simmons-Boardman Publishing. *Locomotive Cyclopedia of American Practice 1922-1941.* Reprinted in *Train Shed Cyclopedia.* Newton K. Gregg Publisher, 1974.

Steinheimer, Richard. *The Electric Way Across the Mountains.* Carbarn Press, 1980.

Swett, Ira. *Montana Trolley III.* Interurbans Magazine, 1970.

Warner, Paul T. *The Chicago Milwaukee & St. Paul Railroad.* Pacific Railway Journal, June 1958.

Wayner, Robert J. *Electric Locomotive Rosters.* Wayner Publications, 1965.

Westcott, Linn H. *Model Railroader Cyclopedia, Volume 1, Steam Locomotives.* Kalmbach Publishing Co., 1960.

Wood, Charles R. *Lines West.* Superior Publishing Co. 1967.

Wood, Charles R. and Dorothy M. *Milwaukee Road West.* Superior Publishing Co., 1972.

Wood, Charles R. *The Northern Pacific: Mainstreet of the Northwest.* Superior Publishing Co., 1968.

ARTICLES:

Buccola, Charles. "Raildates." *Railroad Model Craftsman.* January 1979.

Cummings, Doug E., Ken L. Douglas, and Dick Will. "The Milwaukee Road Electrics Roster." *Extra 2000 South.* January-February 1972.

Gaynor, Joseph N. "Comparative Operating Results of Steam, Diesel-Electric, and Electric Motive Power on the Great Northern Railway Electrification.". *AIEE Transactions,* Volume 67, 1948.

Headlights Magazine. "Those Russian Locomotives." June 1967.

Jones, Vane. "St. Paul Locomotive Tested at Erie." *Electric Railway Journal,* 1919. Reprinted in *Traction Heritage*, Volume 5, Number 3, 1972.

Kelley, W.E. "Historical Summary, Performance and Future of Penn Central Company Railroad Electrification." *Conference on Performance of Electrified Railroads.* 1968.

LeMassena, Robert A. "The Big Engines." *Trains Magazine.* June 1968.

Linebaugh, J.J. "Power Limiting and Indicating System on the C.M. & St. P. Rwy." *General Electric Review.* 1920.

Martin, Thomas M.C. "We Should Have Electrified 15 Years Ago!" *Trains Magazine.* April 1962.

The Milwaukee Road Magazine. "Power and Performance. The Story of the Milwaukee Road's Diesel Engine Fleet." August 1974.

Moe, Glenn S. "South Shore Line." *Railroad Magazine.* February 1953.

Railway Gazette. "Pennsylvania Railroad Improves Its Electric Locomotives." Date not recorded.

Reed, Brian. "The Hiawathas, Locomotive Profile #26." *Profile Publications, Ltd.* 1972.

Wilkerson. Bill. "The Mighty S-2." *The Harlowton Times-Clarion.* 1-7-82.

Wylie, Laurence. *Conference Paper.* "Improvements In The Design, Maintenance, and Operation of the Milwaukee Road Electrification." American Institute of Electrical Engineers. January 30-February 3, 1956.

REPORTS:

Locomotive Products Department Transportation Systems Business Division, General Electric Co. *Proposal for Completion of the Electrified Railway Operation of the Chicago Milwaukee St. Paul & Pacific Railroad Co.* March 20, 1972.

Office of the Division Engineer. *Request for Authority For Expenditure. Retire Electrification Facilities at Falls Yard, Valeria Way and Great Falls.* 9-22-37.

Wylie, Laurence, *Comparative Operating Costs: Electric vs. Diesel.* Coast Division. 1949.

Wylie, Laurence. *Summary Comparison of Diesel vs. Electric Operation for Freight Service—Harlowton—Tacoma.* January 18, 1968.

MANUALS:

Class EF-1 Locomotive Instruction Book. Revised April 2, 1927.

Description and Preliminary Operating Instructions for Class EP-1 Electric Locomotives. February 18, 1953.

Instruction Book. Class EP-2 Locomotives Bipolar Gearless Type for Passenger Service. Revised July 1, 1928.

Instructions for Substation Operation Rocky Mountain, Missoula, and Coast Divisions. August 1922.

Operating Instrutions for Electric Freight Locomotives Classes EF-1, EF-2 and EF-3. April 12, 1950.

Operating Instructions for Electric Locomotives Class EF-4. August 4, 1950.

Rules and Regulations Governing the Operation of the Westinghouse and Electric Passenger Locomotives Class EP-3. Revised September 1, 1926.

Safety Rules Governing Electrification Line Crews and Signal Maintainers When Working on Electrification Poles Wires. Effective March 15, 1931.

Special Rules and Instructions Covering Electrical Operation. Revised June 1958.

Voltage Regulation Instruction Coast Division. August 1959.

INDEX

Alberton 7, 16, 44, 47
Avery 5, 13, 15, 20, 22, 41, 57, 60, 151, 174, 179, 190, 200, 227, 231, 262, 263
Bipolars 86-101
Butte 6, 14, 19, 59, 80, 275
Catenary (trolley) 154-158, 304, 305
Chicago 7
Classification of Wheel Arrangements 279, 280
Cle Elum 7
De-electrification 168, 217
Deer Lodge 5, 10, 22, 23, 24, 68, 84, 85, 97, 107, 262, 269, 277
Diesel Demonstrators 199
Diesel Locomotive Roster 209
Direct Current Drive 281-283
Economics and Company Policy 210-224
Electric Locomotive Specifications:
  Class Designations 290
  Horsepower 291, 292, 294
  Maximum Speeds 291
  Performance Statistics 291
  Tonnage Ratings 295-297
  Tractive Effort 293, 294
Electric Rosters 299-301
Electrical Distribution System 154-169, 306, 307
Electrics Versus Diesels 48, 49, 212-220
Fairbanks-Morse 203
Gap (The Idaho Division) 170-209
GE Motors 33-77
  E22 and E23 69-76
  Freight Motors 38-66
  Passenger Motors 36-37
General Electric Diesels 204
Harlowton 7, 62, 66, 67, 260
Harlowton Switchers 77, 84
Helper Districts 7, 30, 31, 32
Idaho Division 170-209
Kirk, Barry 163
Little Joe Era Electrics 122, 123
Little Joes 118-145
  Freight Joes 130-145
  Passenger Joes 125-129
Locomotive Frames 281
Malden 5, 22, 208
Milwaukee Ski Bowl 3, 28, 90, 150
Milwaukee 7, 35
Missoula 5
Oil Burning Steam Engines 172-177
Othello 3, 58, 173, 266
Pacific Extension 1
Pantographs 288, 289
Regenerative Braking 287
Rotary Snowplows 29, 146-153, 235
Route Maps 2, 4, 6, 223, 224
Seattle 3, 19, 44, 100, 105, 271
Slobber-stacks and Straight Eights 180-185
Spokane 3, 21, 183, 191, 201
Steam Engines
  131 & 132 (4-6-4) 191-193
  C Engines (2-8-0) 189, 190
  F Engines (4-6-2) 186, 187
  I-5 (0-6-0) 189
  K Engines (2-6-2) 189, 190
  L Engines (2-8-2) 188
  N Class (2-6-6-2) 180-185
  S Engines (4-8-4) 194-198
Steam Locomotive Roster 209
Steeplecabs 78-82
Substation Locations 166
Substations 155, 158, 159, 164-167, 211, 302, 303
Tacoma 1, 17, 18, 24, 25, 26, 181, 211
Tales of the Rails
  A Blue Flash at Auburn—Farrow 252
  A Hopper Load of Trouble—Lintz 239, 240
  Drawbars and Garbage Cans—Gratz 235
  E 78 West—Holley 226-232
  Holes in the Roundhouse Wall—Stevenson 241
  Lightning Bolts and Arc Flashes—Kirk 241
  Memories of the Steam Division—Edmunds 242-244
  Orphan Annie—Wilkerson 255-257
  Peak Load on the Motor Division—Chester 245-247
  Riding the Beverly Helper—Dietrich 248-251
  Shoemaker at the Throttle—Edwards 253-254
  Snow in the Passing Track—Gratz 233, 234
  The Runaway—Hansen 236-238
Three Forks 7, 261, 267
Tide Flats 1
Traction Motor Controls 284-286
Trolley 154-158, 304, 305
Trolley Crews 156, 161, 162, 163
Troubleshooters 156, 157
Types of Locomotives 279
Types of Traction Motors and Drive Systems 282, 283
Westinghouse Motors 102-117
Wylie, Laurence 120, 221